INSTANT
PHYSICS

THIS IS A WELBECK BOOK

First published in 2021 by Welbeck,
an imprint of Welbeck Non-Fiction Limited,
part of the Welbeck Publishing Group
Based in London and Sydney.
www.welbeckpublishing.com

Design © Welbeck Non-Fiction Limited 2021
Text copyright © Welbeck Non-Fiction Limited 2021

ISBN 978-1-78739-417-9

Printed in Dubai

10 9 8 7 6 5 4 3 2

The images in this publication are reproduced from thenounproject.com with the exception of
the following pages 24 vchal/Shutterstock, 33, 35 & 45 VecotVectormine/Shutterstock, 47 Fouad A.
Saad/Shutterstock, 52 Julie Deshales/Shutterstock, 67 top Membeth via Wikimedia Commons, 67
bottom Moodswinger via Wikimedia Commons, 69 elenabsi/Shutterstock, 76 SiriusB via Wikimedia
Commons, 102 Udaix/Fouad A. Saad/Shutterstock, 103 Fouad A. Saad/Shutterstock, 116 Designua/
Shutterstock, 124 Vergara Pina Fernando Wla/Shutterstock, 126 Fastfission via Wikimedia Commons,
134 Rhoeo/Shutterstock, 138 Wikimedia Commons, 152 Master_Andrii/Shutterstock, 166 Designua/
Shutterstock, 167 NASA

INSTANT
PHYSICS

KEY THINKERS, THEORIES, DISCOVERIES AND CONCEPTS EXPLAINED ON A SINGLE PAGE

GILES SPARROW

WELBECK

CONTENTS

WAVES

THERMODYNAMICS

ELECTRICITY & MAGNETISM

ATOMS & RADIOACTIVITY

QUANTUM PHYSICS

PARTICLE PHYSICS

RELATIVITY & COSMOLOGY

INTRODUCTION

The story of science is the journey of humanity itself. As a species we've always had an urge to discover and understand the world better. For our earliest ancestors this expressed itself through their exploration and population of the world, and the development of simple technologies that allowed them to shape it and improve their lives. But almost as soon as people began to adopt settled lifestyles, we started to ask deeper questions about why the world works in the way it does.

Why does the Sun rise and set, and where does it go when it's out of sight? What are the other lights that spin across the night sky? Why do the seasons change, and can we predict them to improve our lives? How can we build better shelters and monuments to keep warm and worship our deities? At first, attempts to answer these questions deferred to supernatural causes, but in Ancient Greece around the middle of the first millennium AD, a new approach emerged. Natural philosophy was based on the assumption that the Universe did not follow the arbitrary whims of the gods, but instead obeyed strict laws that governed its various aspects. At first, however, ideas about what these laws might be were based not entirely on the reality of the world around us, but also on best guesses at what forms and behaviours might be most appropriate, "perfect", pleasing to nature, or in line with religious teachings.

It was not until late medieval times that thinkers in Europe began to break away from these old ways of thinking, developing a new approach in which theories and hypotheses were developed more strictly from observation, and tested with experiments. The application of mathematics to understanding various aspects of the world – championed by the Italian polymath Galileo Galilei – gave rise eventually to something more like modern science.

In this new age of discovery and experimentation, the field we now call physics rapidly rose to prominence – in part because its concern with measurable phenomena such as force, mass, and motion lent itself to the mathematical approach. Yet it's only recently that we've come to recognize that the scope of physics is not confined to the orbits of planets, the fall of apples, and the motion of projectiles.

Since the seventeenth century, successive breakthroughs have shown how physics is essential to everything in nature. The force of gravity has been joined by three other fundamental forces that explain how matter interacts on a huge range of scales from the very large to the very small. The structure of matter, meanwhile, has given up its secrets – division into atoms of many different elements, and below these, into subatomic particles that are governed by the strange laws of quantum mechanics. Again and again, however, the behaviour of matter, in fields from cosmology and chemistry to biology and electronics, keeps coming back to the same physical concerns of forces, motion, mass, and energy.

The story of physics is therefore far wider than that which many of us learned in our school textbooks. In some of its most recent chapters, physicists have turned their attention to undetectable forms of matter that account for the vast majority of mass in the universe and drive its surprising expansion; they've learned to forge and manipulate strange bonds between subatomic particles on the quantum scale; and they've uncovered links between fundamental forces that offer tantalizing hints that a "Theory of Everything" might one day be possible.

At its heart, science – and physics especially – is all about assessing what we know about the way in which the world and the universe around us behave, asking the right questions, thinking up possible answers, and testing them. It's a combination of lateral thinking and spotting the obvious but fundamental points that others have overlooked – and that's a thread that runs through its history from Galileo to the present.

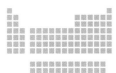  

To an outsider, however, all this can seem baffling – and that's where *Instant Physics* comes in. Across 160 pages, this book covers everything from the very basic equations you've forgotten learning at school, to the cutting-edge theories and discoveries of the latest research.

The book is divided into ten sections. After a broad look at the story of physics from Ancient Greece to the present day, each covers one of the major fields within physics, in rough order of historical development. Dip in at any page and you'll hopefully learn something you didn't know, but if you want to follow as much as possible, start at the start, with the sections on Mechanics (those fundamental forces and rules on which everything else is founded) and Matter. Helpful diagrams are peppered throughout the book, along with plenty of useful equations if you've a head for maths (though you should still be able to follow the ideas if you haven't).

Physics is the science of everything, and everyone should have a grasp of at least the basics – so dive in and start discovering!

FIRST PHYSICISTS

The word physics comes from the Greek word for nature, physis. Physics is the study of nature – specifically of the non-living processes involving matter, motion, force, and energy. The Ancient Greeks were the first to attempt to explain such phenomena.

Thales of Miletus (c.624–c.548 BCE)
The **first scientific philosopher**, Thales **rejected** the idea of **supernatural forces governing the world** and sought to explain its properties through **natural materials** and **forces**. He was the first to investigate phenomena such as **static electricity**, and **believed that everything in the Universe** was **derived from water**.

Pythagoras (c.570–c.495 BCE)
Famous for his **mathematical theorem regarding triangles**, Pythagoras's teachings inspired an **entire school of his followers** to pursue **mathematical explanations of natural patterns** such as the **harmonics** of **musical notes**.

Anaximander (c.610–c.546 BCE)
A **pupil of Thales**, Anaximander believed that **everything stemmed** from a **concept** called the *apeiron* or **"indefinite"**, which could give rise to **space**, **time**, and different forms of **matter**. He attributed the **laws of physics** and the **variety of matter** to the **separation of opposing properties** (**hot/cold, dry/wet** and so on).

Heraclitus (c.535–c.475 BCE)
Heraclitus is famous for his **conception of reality** as a **process of endless change**, encapsulated in the phrase **"one never steps into the same river twice"**. He considered **fire** to be the **primordial element** from which **everything else stemmed**, and believed that **all things** contained a *unity of opposites*.

Empedocles (c.494–c.434 BCE)
Drawing on ideas from **earlier philosophers**, Empedocles produced an influential **"Theory of Everything"**. Four **elements** – **earth**, **air**, **fire**, and **water** – are **mixed** or **separated** to create **different forms of matter** by **interaction** with **attractive** and **repulsive forces** known as **love** and **strife**.

Democritus (c.460–c.370 BCE)
An **early proponent of atomic theory**, Democritus argued that **all matter** was **constructed of tiny indivisible units**, with **empty space between them**. He believed the **properties** of **different forms of matter** arose from the **shapes of atoms**, but his **concept of void** was **widely rejected**.

ARISTOTLE AND THE ELEMENTS

Among the greatest of all ancient philosophers, Aristotle (384–322 BCE) was also the one with most to say about the workings of the Universe. His views were hugely influential and were rarely challenged for almost two thousand years.

COSMOLOGY AND PHYSICS

Contrasting the **apparently perfect cycles of motion** in the **heavens** with the **unpredictable changing nature of Earth**, Aristotle argued for a **fundamental division** of the **Universe**:

- Everything on **Earth** consists of the **four elements**: earth, air, fire, and water.
- The **Moon** and **everything beyond it** are composed of an **incorruptible fifth element: aether.**

Each element had a **natural place** in the **Universe**: earth belongs at the **centre** of the Universe, with **water above it**. **Air tended to rise upwards**, fire more so, while aether was **actively repelled** from the **corruption** of the **other elements.**

The earthly elements had **pairs of properties**, which they **shared with their neighbours** – altering one property changed one element to another.

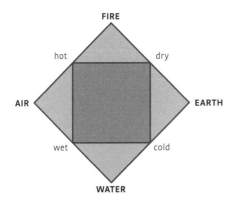

ARISTOTLE'S MOTION

For Aristotle there were **two forms of motion**:

- **Natural motion** – movement **towards** an object's **proper place** depending on its **elemental composition**.
- **Unnatural** or **violent motion** – movement **away** from an object's "proper" place, which is **gradually overcome by natural motion.**

Aristotle had **no need for gravity** – the **tendency of objects** to get as **close as possible** to their **natural place** in the (Earth-centred) Universe explained **all natural motion**.

He believed this tendency was **directly linked to an object's weight**, so that **heavier objects must fall more rapidly than lighter ones** containing less "**earth**".

Dense Fluid

Heavier ball falls faster

$v \propto W$

Violent motion, however, required the **continuous application of a force**, and the **speed of an object in violent motion** was **proportional** to the **strength of this force**.

ARISTOTLE'S METHOD

Aristotle claimed the **best way to learn about the Universe** was to form **general models based on a collection of examples**. This **inductive method** sounds quite like **modern science**, but Aristotle **did not then test his theories through experiments**. He was also **reluctant to abandon many aspects of received philosophical wisdom**.

ARCHIMEDES

Greek philosopher and mathematician Archimedes (c.287–c.212 BCE) is also regarded as the greatest engineer of the ancient world. He was the first to outline the principles behind many of the devices now known as simple machines.

ARCHIMEDES' MACHINES

Archimedes lived and worked in the **Greek colony of Syracuse**, on the island of Sicily, where he became **renowned** as a **mathematician** and **inventor** (though **many accounts of his inventions come from later writers**).

His sophisticated understanding of **forces** allowed him to **devise** a number of **machines** that **multiplied applied force** in order to **make mechanical tasks easier**. These included:

- **Archimedes' screw**: A revolving screw fitted inside a **cylinder**, used to **raise water** from a **lower level** to a **higher one**.

- **The Claw of Archimedes**: A **crane** with a **lever-like arm**, a **grappling hook suspended from the end**, and a **system of pulleys** that allowed a **small number of people** to **lift a heavy load** – specifically, **overturning an attacking ship**.

ARCHIMEDES' PRINCIPLE

Of course, Archimedes is **most famous** for the principle that a **submerged object experiences** an **upward buoyant force equivalent to** the **weight of water** or **other fluid** it **displaces**.

The tale of Archimedes being asked to **measure** the **composition of a crown** and solving this in the **bath** is probably **apocryphal**, but he did devise a **method for accomplishing such tasks** in his treatise *On Floating Bodies*. This involved **balancing two masses on a scale** and then **immersing** the **entire scale** in **water** to **measure the buoyant force on each**.

ARCHIMEDES GOES TO WAR

When Syracuse was **besieged by Roman forces** in 214 BCE, Archimedes turned his attention to the **defence of his home city**. He supposedly devised a variety of **war machines** and **secret weapons**, including not only the Claw of Archimedes, but also a **heat ray** that used **curved mirrors** to **focus sunlight on Roman ships** and **set them alight**, and a **greatly improved catapult**.

EARTH AND THE HEAVENS

Various theories about the relationship between Earth and the celestial bodies in the day and night skies were proposed in the ancient world, but the assumption that Earth was the centre of everything was widely held.

EARLY COSMOLOGIES

- **Philolaus** (c.470–c.385 BCE): Argued for a **spherical Earth** with the **inhabited world confined to one side**. The **visible celestial objects**, as well as **Earth itself** and a **balancing counter-Earth, orbited** a "**central fire**", **permanently out of sight** on the **far side of the world**.

- **Eudoxus** (c.390–c.337 BCE): Introduced the idea of **celestial spheres. Each planet** sits on the **innermost** of a **set of nested, transparent spheres rotating at different rates. Interactions** between these **different rotations** explain the **wandering motion** of the **planets** compared to the **outermost sphere** of **fixed stars**.

- **Aristarchus** (c.310–c.230 BCE): Proposed a **Universe** with the **Sun**, rather than the **Earth**, at its **centre**. He successfully showed that the **Sun** was **much larger** and **more distant** than the **Moon**, but the **lack of direct evidence** that **Earth was in motion** meant his idea **struggled to win support**.

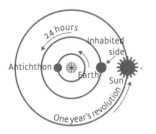

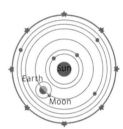

PTOLEMY AND THE ALMAGEST

Egyptian-Greek astronomer **Ptolemy** (c.100–c.170 CE) produced the **most detailed Earth-centred cosmology** in a book known as the *Almagest*:

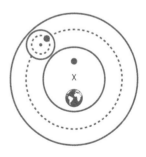

- He **maintained Aristotle's concepts** of **crystal spheres** and **uniform circular motion**.
- To this, he added **epicycles** – **smaller spheres** that allowed the **planets to circle back and forth** as they were **generally carried in one direction** by the **main sphere**.
- While the **planetary spheres remained circular**, their **centres did not necessary line up** with the **centre of the Earth**.

Ptolemy's theory appeared at first to **solve the mismatch** between **theory** and **observations of planetary motion**.

It was **hugely influential** and became **accepted wisdom** for **more than a thousand years**.

However, as time went on, Ptolemy's model proved **incapable of accurately predicting planetary motions**, with **later astronomers adding more and more epicycles** to the system in an effort to fix it.

ORIGINS

15

ISLAMIC ADVANCES

While the late Roman Empire saw the decline of fresh thinking in natural philosophy and an increasing deference to received wisdom, the Islamic Golden Age saw a resurgence of new ideas and approaches.

PRESERVATION AND INNOVATION

The late 700s saw the establishment of the **House of Wisdom** in **Baghdad**, and the beginnings of a **translation effort** that aimed to make **surviving texts** from the **Classical world**, gathered as **Islam** had spread across the **abandoned territories** of the **fallen Roman Empire** in the previous century, **available in Arabic**.

At the same time, scholars did not merely **copy manuscripts** but also aimed to **provide commentary** and **improvement** upon them, sometimes by **directly testing their ideas** with **experiments**.

This **Islamic Golden Age** lasted well into the second millennium and saw many **new ideas in physics**:

- **Ibn al-Haytham** (c.965–c.1040) expressed **doubts about Ptolemy's cosmology** and suggested that **celestial objects** were **subject** to the **same forces** as **those on Earth**.
- **Ibn Bäjja** (c.1085–1138) suggested that when **one object applies a force to another**, the **first object** itself **experiences a reaction force** – an **early form of Newton's third law**.
- Around 1125, **al-Khazini** suggested that a **force of gravity acts on all objects** and **varies in strength** depending on their **distance** from the **centre of the Universe** (namely **Earth**).

LIGHT AND OPTICS

Islamic philosophers made great strides in their investigations of the **nature of light** and the **behaviour of lenses** and other **optical devices**.

Many **Greek thinkers** (including **Ptolemy** and the influential physician **Galen**) had believed in an **emission theory of vision**: light originated as **streams of particles** emitted from the eye, whose **reflections returned sensory information to an observer**.

The first comprehensive *intromission* **theory of vision** was formulated by **Ibn al-Haytham**. He carried out **experiments** with **lenses** and **mirrors** to show that **light travelled in straight lines**, and realized this somehow **produced an image inside the eye**, although the details eluded him.

Ibn Sahl (c.940–1000), meanwhile, identified the **law of refracted light** widely known as **Snell's Law** (see p.71), and used this to **work out** the **ideal shapes for magnifying lenses**.

MEDIEVAL INNOVATIONS

Medieval Europe inherited many ideas from Classical times, filtered through the teachings of the Christian Church. As these mixed with lost texts reintroduced from the Islamic world, they inspired some surprisingly fresh thinking.

THE NOT-SO DARK AGES

The traditional view of the medieval world is that a hierarchical imposition of dusty received wisdom from the **Church** and **authorities** such as Aristotle **stifled new thinking, especially in the sciences**. The reality, however, is rather different, with some **surprisingly insightful breakthroughs**:

- In the sixth century, Byzantine scholar **John Philoponus** had already **questioned Aristotle's suggestion** that **bodies of different masses fall at different rates**. He also argued that **forces need not act continuously on a projectile** to **keep it in violent motion**, but could instead **impress an** *impulse* that **declined over time**.
- **Jean Buridan** (c.1301–c.1362) took the ideas of Philoponus (already processed through several **Arabic philosophers**) and fashioned a **theory of impetus**, an **imparted property** that **did not decline**

spontaneously, but was instead **drained by forces** such as **gravity** and **air resistance**.

- Buridan's pupil **Albert of Saxony** (c.1320–90) showed that the **velocity of a falling body** was **proportional** to the **square of the time elapsed since release** or the **distance already fallen**.
- In the mid-fourteenth century, a group of scholars at **Merton College, Oxford**, applied **mathematics** and **logic** to various questions in "**natural philosophy**". One of their key results was a **rule of uniform acceleration**.
- **Nicole Oresme** (c.1325–82) **proved the Merton rule** by **drawing graphs** – an approach that was hugely influential. He also dabbled with the idea of a **Sun-centred Universe**, and **criticized the thinking behind astrology**.

MERTON RULE – NICOLE ORESME

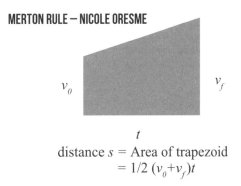

$$\text{distance } s = \text{Area of trapezoid}$$
$$= 1/2\ (v_0 + v_f)t$$

CHANGING PERSPECTIVES

From the time of Aristotle, **mechanics** was **held back** by the **search for a prime mover**, the supposed **driver of all motion** in the **Universe**. In the late Middle Ages a **new field** of **kinematics**, which focused on **describing motion without worrying about its causes**, led to significant advances. However, kinematics was still rooted in **logic** and **mathematics** – the idea of **experimenting** to **test hypotheses** had **not yet taken hold**.

THE COPERNICAN REVOLUTION

In 1543, Polish priest Nicolaus Copernicus published a book suggesting that the Sun, rather than the Earth, was the centre of the Universe. The revolution he began was a landmark on the road to modern science.

ORBITING THE SUN?

Copernicus's suggestion that **Earth orbited the Sun** was based on **observations** of the **planets' movement** in the sky:

- **Mercury** and **Venus** are **only ever seen close to the Sun** in **mornings** and **evenings**.
- **Mars** moves **westward** around the sky but makes **large retrograde loops** where it **reverses** for a **few months at a time**.

- **Jupiter** and **Saturn** progress more **slowly** around the sky with **smaller retrograde loops**.

As the **accuracy of measurements improved**, **philosophers** struggled to **modify** the **Earth-centred model** of the **Universe** to **match with reality**.

In his *On the Revolutions of the Heavenly Spheres*, Copernicus argued that a **Sun-centred model** could **explain these motions** just as well. For instance, the **retrograde motion of Mars** could be an effect of our **changing point of view** as planet Earth "overtakes it".

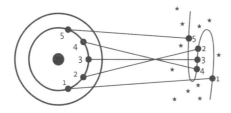

However, Copernicus **undermined his model** by imagining **perfectly circular, uniform orbits**, which offered **no improvement** for **explaining** the **observed motions of the planets**.

Decades later, **Galileo** provided **irrefutable proof** of the **Sun-centred theory** through **observations with his telescope**, while **Johannes Kepler** provided a **mathematical model** that **finally meshed with observation**.

LEARNING OUR PLACE

The **Copernican revolution** was just the **first of many** that have successively **overturned ideas** about the **special place of humanity** and **Earth** in the **Universe**:

- **1838** Friedrich Bessel measures the distance to another star, revealing the true scale of interstellar space and the Milky Way

- **1859** The Darwin/Wallace theory of evolution explains how humans could arise by a process of natural change rather than divine creation

- **1925** Edwin Hubble proves that the Milky Way galaxy is just one among countless millions

- **1990s** The discovery of the first exoplanets around other stars overturns the idea that Earth's solar system might make it special

- **2000s** Evidence increasingly suggests that our Universe is one of many in a potentially infinite multiverse

GALILEO'S EXPERIMENTS

Italian mathematician Galileo Galilei (1564–1642) is most famous for his telescopic observations and his battle with the Catholic Church, but he also played a key role in establishing a recognizably modern approach to physics.

MATHEMATICS AND MEASUREMENT

As **professor of mathematics** at **Pisa** and later **Padua**, Galileo pioneered an **approach to natural philosophy** that **combined experimentation with mathematical modelling**. He applied this to a wide variety of **physical phenomena**, including:

- Finding a **mathematical relationship** between the **length of a pendulum** and the **period of its oscillations**.

- Demonstrating that **bodies fall** with the **same acceleration regardless of mass**, in **direct contradiction of Aristotle**.

- Linking the **pitch of a sound** to the **frequency of its waves**.

- Developing the **thermoscope**, a predecessor of the **thermometer**.

- Establishing the **principle of relativity** – that the **laws of physics** must be the **same for any system in constant straight-line motion**, regardless of its **speed** or **motion relative to another frame of reference**.

STUDYING THE UNIVERSE

In 1609, after hearing reports of a **new invention** from the **Netherlands**, Galileo built his **first telescope**. His experimental approach allowed him to rapidly **improve his original design**, and his **observations** of the heavens revealed important discoveries that **undermined the Aristotelean and Ptolemaic view of the Universe**. These included:

- Four **satellites** circling **Jupiter**
- **Moon-like phases** on **Venus**
- **Craters** and **mountains** on the surface of the **Moon**
- Changing **spots** on the **Sun**

Galileo's discoveries turned him into a **staunch advocate** of the **Copernican, Sun-centred view of the Universe** – an idea that **went against** the teaching of the **Catholic Church**. **Powerful friends in the Church** initially suggested a **compromise** position, but Galileo's insistence on the truth of the Copernican system ultimately led to his fall from grace and **house arrest** from 1633.

KEPLER'S LAWS

Through a combination of observation and hypothesis, German mathematician Johannes Kepler (1571–1630) identified laws of planetary orbits that would ultimately provide the key to unlocking the mysteries of motion in all its forms.

KEPLER AND TYCHO

Kepler's discoveries were founded on the **precise observations** of Danish astronomer **Tycho Brahe** (1546–1601). Working **before the telescope**, Tycho used large instruments called **mural quadrants** to precisely measure the **positions of planets**:

· The **Quadrant** is mounted on a wall **aligned north– south**, so objects are **observed at their highest point** in the sky.
· An **Alidade** bar is fixed at the **uppermost southern corner** and **swings through a 90° arc**.
· A **sight** allows **precise alignment to stars** along the alidade, revealing their **elevation** in the sky.
· **Precise timing** of **meridian transit** – when an object passes **due south** – reveals its **right ascension** relative to other stars.

Kepler worked as **Tycho's assistant** from 1600, succeeding to his post as **Court Astronomer to Rudolf II of Prague** after his death, and inheriting a **wealth of accurate observations** – most significantly those tracking the **path of Mars**.

ELLIPTICAL ORBITS

In his 1609 *Astronomia Nova*, Kepler described the **orbits of planets** by **abandoning** the **long-held belief in circular orbits**, in favour of **ellipses** (**circles** that are **stretched along one axis**). Kepler's **three laws of planetary motion** state that:

· **Any planet's orbit** is an **ellipse**, with the **Sun** at one of the two *foci*.
· An **imaginary line linking the Sun to a planet** sweeps out **equal areas** in **equal times** (so the **planet moves faster** when it is **closer to the Sun** and **slower** when it is **further away**).
· The **square** of the **orbital period** increases in **proportion to the cube of the orbit's semi-major axis** (half the length of its longest axis).

Kepler's laws **accurately described planetary movements** and **provided tools for predicting future motions**. However, it was not until the 1680s that **Isaac Newton** would explain ***why*** these particular **relationships held true**.

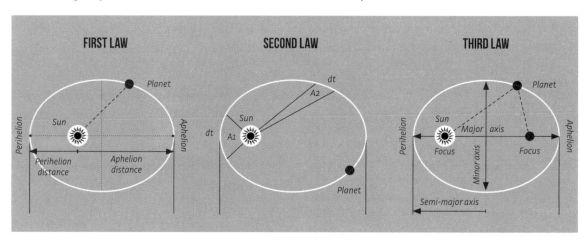

FIRST LAW · SECOND LAW · THIRD LAW

ISAAC NEWTON

Isaac Newton (1642–1727) transformed mathematics and optics, as well as constructing a hugely important theory of motion and gravitation based on Kepler's laws and his own observations of falling bodies on Earth.

MOTION IN HEAVENS AND EARTH

Newton's **great theoretical breakthrough** (as early as the 1660s) was to understand that the **forces pulling a falling apple towards Earth** are the **same as those holding a planet on its orbit**:

- Gravity draws objects directly towards the centre of massive objects.

- The gravity is in proportion to the attracting mass.

In the 1670s he developed the **mathematics of calculus** – a method for **understanding processes** by looking at **how they behave** with the **tiniest variations**.

In the 1680s, encouraged by **Edmond Halley** of comet fame, Newton applied his methods to **Kepler's laws of planetary motion**, showing how the **patterns of elliptical motion** they described could be caused by an **attractive force towards the Sun** that **diminished in proportion to the square of a planet's distance**.

In his 1687 *Principia*, Newton outlined an **entire system of mechanics** that extended from the **commonplace** to the **cosmic**, including his **three laws of motion** (some of which had been established in late medieval times) and an **equation for universal gravitation** linking the **masses of bodies** with the **distance between them** and showing why **acceleration due to gravity** is **identical for all falling bodies**.

NEWTON'S OPTICS

Newton's other great area of scientific study was the **investigation of light**, summarized in his *Opticks* of 1704:

- Newton showed that **white light** from the **Sun** is actually a **mix of colours** that can be **split by a prism** and also **recombined**.

- He designed the **first reflecting** or **mirror-based telescope**, forerunner of today's giant observatories.

- Newton argued that light was **composed of corpuscles – tiny particles** much more "**subtle**" than those of **matter**. However, in the early nineteenth century this was **overturned** in favour of a **wave theory**.

THE NEWTONIAN LEGACY

Isaac Newton's breakthroughs in the fields of mechanics and optics ensured a long-lasting legacy. For much of the eighteenth century, physicists were largely concerned with the practical applications of his fundamental laws.

PUTTING NEWTONIAN PHYSICS TO WORK

Newton's laws described situations involving **point masses** and **bodies moving freely in space** – ideal for problems such as **simple planetary motion**, but **harder to apply to everyday forms of motion. Key breakthroughs** in the **practical application of Newtonian mechanics** included:

1732 Daniel Bernoulli applies Newton's laws to each segment of an oscillating string, identifying a form of periodic movement called simple harmonic motion

1736 Leonhard Euler divides the apparently random motion of ships and other rigid bodies into rotational and translational components

1740s Émilie du Châtelet identifies an equation for describing *vis viva* (akin to modern kinetic energy), and argues that the total energy in a system is always conserved even if it is transferred between different forms

1798 Joseph-Louis Lagrange puts forward his analytical mechanics – a set of mathematical tools for modelling the behaviour of systems subject to a constraint in one or more dimensions

Nevertheless, **various challenges remained** – not least the **long-term unpredictability of systems involving more than two bodies** and **puzzling changes** to the **orbit of Mercury** (later explained through general relativity).

PRINCIPLE OF LEAST ACTION

From 1827, Irish mathematician **William Rowan Hamilton** identified a **fundamental principle underlying mechanical interactions** – namely that **objects tend to move along the path that requires least energy**. Thus, a **huge range of mechanical problems could be solved** simply by finding the **turning points** on a **graph describing motion**.

By 1833, Hamilton had formulated an **entirely new approach to mechanics**, using **equations** that described a **system's evolution over time** in terms of **generalized coordinates**, the **total energy** (today denoted $\mathcal{H}$ and known as the "**Hamiltonian**") and the **balance of kinetic and potential energies**.

THE AGE OF ELECTROMAGNETISM

The nineteenth century saw huge breakthroughs in the understanding of electricity and magnetism, ultimately leading to the realization that these two forces, along with light itself, were all aspects of a single phenomenon.

ELECTRICAL BREAKTHROUGHS

Although **electricity** had been **known since ancient times**, **experiments** into its true nature **only became practical** after **Alessandro Volta's invention** of the **voltaic pile** (the **first battery capable of producing continuous current**) in 1800:

1820 Hans Christian Ørsted notices that changing electric current can deflect a magnet needle – the first evidence for a unified force of electromagnetism

1820 André-Marie Ampère discovers the force generated between two parallel conducting wires

1821 Michael Faraday invents the electric motor, which uses the interaction between electric current and magnets to power mechanical movement

1827 Georg Ohm describes the relationship between electromotive force (voltage), current, and resistance within circuits

1831 Faraday discovers electromagnetic induction, the generation of electric current by moving magnetic fields, allowing him to build the first electrical generator

1845 Faraday discovers that the orientation of polarized light waves is rotated as they pass through a magnetic field

ELECTROMAGNETIC WAVES

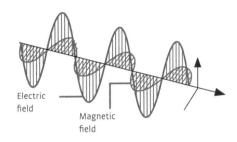

Electric field

Magnetic field

Between 1861 and 1862, Scots mathematician **James Clerk Maxwell** attempted to put the **effects of electromagnetic fields** on a **sound theoretical footing** using a **model of rotating vortices** and **fixed particles**. He discovered that **changes to such a field** would **propagate** at an **approximate speed** of **311,000 kilometres per second** – suspiciously **close** to **estimates** of the **speed of light**.

Faraday had already found evidence that **light** was **affected by electromagnetic fields**, and Maxwell was able to explain just how the **Faraday rotation** came about. In 1864, he published *A Dynamical Theory of the Electromagnetic Field*, in which he described light itself as a **moving wave of electrical** and **magnetic disturbances** at **right angles** to each other.

In 1886, **Heinrich Hertz** proved Maxwell's theory beyond doubt with his **discovery of radio waves** – with **wavelengths much longer than visible light**.

ALBERT EINSTEIN

Though best known today for his theory of relativity, Albert Einstein's discoveries and predictions across a range of fields helped lay the foundations of physics in the twentieth century and beyond.

EINSTEIN'S ANNUS MIRABILIS

In 1905, the unknown Einstein published **four ground-breaking papers** that addressed a series of **major problems** that lingered in the physics of the late nineteenth century:

- **Direct evidence for atoms**: Einstein showed how Brownian motion (the **apparently random jiggling of small particles in fluids**) could be explained as the effect of **interaction** with **countless tiny, invisible atoms** and **molecules**.
- **Particles of light**: To explain why **certain metals produce electric current** when **illuminated by short (but not long) wavelengths of light**, Einstein argued that the light was **delivered** in **tiny packets** or **quanta**. The suggestion that light has both **wave** and **particle aspects** lay at the heart of later **quantum physics**.
- **The fixed speed of light**: Since experiments had consistently **failed to detect changes** in the **speed of light** due to the **relative motions** of **source** and **observer**, Einstein made an assumption that light's speed was truly **fixed**. His **theory of special relativity** showed that, although **Newtonian models remain correct in most situations**, strange effects arise when two non-accelerating frames of reference move at very high relative speeds.
- **Mass–energy equivalence**: Building on special relativity, Einstein now considered the **energy of bodies at speeds approaching** the **speed of light**. This led him to conclude that **mass** and **energy** are **equivalent**, summarized in his famous equation $E = mc^2$.

GENERAL RELATIVITY AND BEYOND

Einstein's 1915 **theory of general relativity** made a **fundamental link** between **acceleration** and **gravitation**. It showed that **strong gravitational fields** could produce **similar effects** to **special relativity**, and that **all of these effects** could be treated as **distortions** of **space** and **time** themselves.

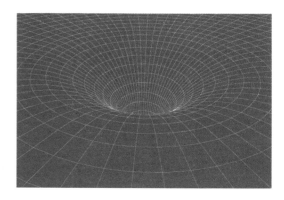

Einstein's later career was devoted to exploring the **implications of special and general relativity**, and to attempts to **unify relativity** with the **very different equations of quantum physics**.

TWENTIETH-CENTURY ADVANCES

The twentieth century saw huge advances in physics at both extremes of scale – the discovery of atomic structure and the uncertainties of the subatomic realm, and a deeper understanding of the true nature of the cosmos.

INSIDE THE ATOM

From 1897 onwards, physicists discovered a **series of subatomic particles** that provided an **explanation** for the **deep structure of matter**, the **bewildering variety of elements** and the **strange phenomenon known as radioactivity**.

In the 1920s, studies of **subatomic particles** showed that on the **smallest scales**, the **properties** of particles **cannot be known with absolute precision** and are **better described** by **wavelike equations**. This opened up an **entirely new field** of science – **quantum mechanics**.

The discovery of the **atomic nucleus** revealed a need for not one, but **two new fundamental forces** in addition to **electromagnetism** and **gravitation** – the **strong** and **weak interactions**.

Since the 1950s, increasingly powerful **particle accelerators** have revealed a **host of new subatomic particles**, fleshing out a so-called **Standard Model** of **how particles and forces interact**. But many unanswered questions still remain.

THE COSMOLOGICAL REVOLUTION

Einstein's 1915 theory of general relativity **reshaped our perception of space and time**, but an improved understanding of the **Universe as a whole** has come from **astronomical observation** since then.

1925 Edwin Hubble found conclusive evidence for galaxies beyond the Milky Way, expanding the scale of the Universe to many billions of light-years

1929 Hubble found evidence that distant galaxies are receding at high speeds, and that the Universe as a whole is expanding

1931 Georges Lemaître suggested that cosmic expansion could be traced back in time to a hot, dense state, far back in time, out of which the Universe was born

1964 Astronomers discovered a faint afterglow from this initial "Big Bang" permeating the modern Universe

1980 Vera Rubin published evidence (from the rotation of spiral galaxies) that 85 per cent of all mass in the Universe takes the form of undetectable dark matter

1999 Astronomers found evidence that cosmic expansion is accelerating rather than slowing down, due to a mysterious force called dark energy

2001 Hubble Space Telescope measurements of cosmic expansion pinned down the date of the Big Bang itself to 13.7 billion years ago

SPEED, VELOCITY, AND MOTION

In physics, the similar-sounding concepts of speed and velocity have distinct meanings – understanding the difference between them, and how their values can change, is key to describing the behaviour of bodies in motion.

SPEED OR VELOCITY?

- **Speed** is simply a **measure of the rate** at which a **moving body changes position over time** – it is measured in units such as **miles per hour** or **metres per second**, and the **direction of motion** is **irrelevant**.

- **Velocity**, however, is the **rate of motion** in a **specific direction** (though one that can be **arbitrarily decided depending on the situation** in order to make calculations easier).

Speed is a **scalar quantity** – one with a **magnitude** or **value**, but **no direction**.
Velocity is a **vector quantity**, with both a **magnitude** and a **direction**.

Velocity is far **more useful** for most **mechanical calculations**, since the **forces** that affect an **object's motion** are likely to **act in a specific direction**.

ACCELERATION

Somewhat confusingly, the term **acceleration** can be used to describe the **rate of change** in either **speed** or **velocity**, **depending on the situation** and the **mechanical system being described**.

Acceleration describes **how much** the **speed** or **velocity changes** in a **given unit of time**, so it's measured in units such as **metres per second per second**.

Deceleration is simply **negative acceleration** that **slows down** an **object's speed** or **velocity** instead of **boosting it**. Physicists use the term acceleration for **both**.

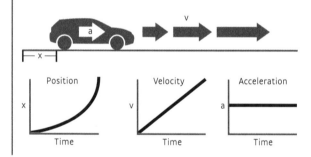

MEASUREMENT UNITS

Rather than simply **abbreviating units of measurement**, scientists use a **mathematical style of notation**. For example:

A **superscripted number** indicates a **power** (**multiplication of the unit with itself**), and the − (**minus) symbol** indicates that the unit is being **used as a divisor** – i.e. it **belongs below/to the right of** a **division line**).

Property	Unit	Abbreviated	Scientific
Speed	kilometres per second	km/s	kms^{-1}
Acceleration	metres per second per second	m/s/s	ms^{-2}

MASS AND WEIGHT

Although the words are often used interchangeably, mass and weight also have very distinct definitions in physics – mass is an inherent property of an object itself, while weight arises from its surrounding environment.

WHAT'S THE DIFFERENCE?

For **non-physicists**, the **distinction** between **mass** and **weight** seems **confusing** and **nit-picky**. Units such as **kilograms**, **pounds** and **stone** that we habitually use to **describe "weight"** are actually **measures of mass**, while **weight itself** is measured in an **entirely different unit** called the **newton**.

Mass is a **direct measure** of the **amount of material in an object** and therefore **how hard it is to move** (sometimes called its **inertia**), and **how strong an attractive force it exerts due to that mass** (its **gravity**).

Weight is properly described as a **measure** of the **force a particular mass exerts or experiences due to the gravity in its environment**.

On the **surface** of the **Earth**, an **object's weight** is the **force** that **causes it to fall "down" towards the centre of the Earth**. This fall may be **arrested** or **prevented** by the presence of a **barrier** or a **counteracting upward/outward force**, but the **downward force** or **weight itself remains**.

INERTIA MATTERS

Outside of a **gravitational field** (or when the **force of gravity** is **neutralized** by **other forces** – as in **orbit**, see p.20), an **object has no weight**, but its **mass remains the same**.

For example, in **weightless conditions** a **bowling ball** and a **balloon** might both **float in mid-air**, but the **bowling ball's greater mass** and **inertia** means it takes **much more effort to move it**.

MECHANICS

FORCES AND MOMENTUM

The concept of force is perhaps the most important idea in physics – and one that crops up on every possible scale and in a wide variety of apparently different situations.

FORCES FOR BEGINNERS

Put simply, a **force** is an **effect** that **produces a change** in an **object's motion**. The **bigger the force**, the **greater the change** (in accordance with **Newton's second law of motion**).

A force can be **exerted** through **direct interactions of objects** such as **collisions**, or by **force fields** that **affect any susceptible objects within their range**. The **most familiar force field** is **gravity**, which **influences all objects with mass**.

Scientists analyse **forces** and **motions** in terms of **systems**. A system is a **small chunk of space** with **any objects** that **it contains**, and **any external force fields** that **act within it**.

Forces are **measured in newtons**, where

1 newton is the force required to accelerate a mass of 1 kg at a rate of 1 ms^{-2}.

WHAT IS MOMENTUM?

Momentum is the **physical property of objects** that is **affected by force**. Crudely stated, it represents the **difficulty of stopping the object's motion in its tracks**. It is **usually represented by the letter p**, and is **calculated by multiplying an object's mass by its velocity**:

$$p = mv.$$

Momentum is measured in **kilogram metres per second** (kg ms^{-1}). Note that its **value depends on velocity, not speed** – so when **objects in a system move in opposite directions**, their **individual momenta have opposite signs**.

This is **important** because the **total momentum in a system is always *conserved*** – **provided no external forces are present**, the sum of the momenta of individual objects remains the same before and after objects interact.

Pool balls offer a **simple example** of **conservation of momentum in action**. When a **fast-moving cue ball packed with momentum hits the pack**, it often **stops dead** while **transferring a little of this momentum to all the other balls**.

FRICTION

Friction is the most common reason why the behaviour of everyday objects doesn't always match the simplicity of basic physics. In almost every aspect of the Universe, it exerts a force that drains away the momentum of moving objects.

WHAT IS FRICTION?

Friction is a **force created by interactions with an object's environment**. It tends to **slow down objects in motion** and **make static objects harder to move than they would otherwise be**. It is found in three main forms:

- **Fluid friction**: created by **collisions with molecules** in a **gas** or **liquid**. **Moving objects experience resistance to motion**, while **individual molecules** or **atoms** in the **fluid** and the **moving object** are **heated** and **gain energy**.

- **Internal friction**: **forces** between **molecules** within **solid bodies** that **resist attempts to deform them**.

- **Dry friction**: **weak chemical bonds** between **solid surfaces**, along with the **physical interlocking** of their **irregularities**, creates **resistance to motion**. Dry friction **behaves differently** depending on whether the surfaces are **moving past each other** (**kinetic friction**) or **standing still** (**static friction**).

COEFFICIENTS OF FRICTION

For **dry friction**, the **coefficient of friction** μ is the **ratio of the friction force** to the **force pressing one surface onto the other**. The **coefficient of static friction** μ_s is **generally higher** than the **coefficient of dynamic friction** μ_k, because there are **more frictional forces to overcome** when **getting an object moving** than in **keeping it moving**.

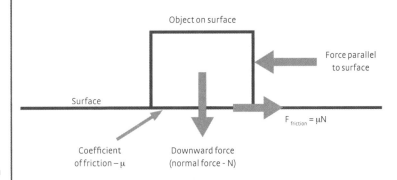

Object on surface

Force parallel to surface

Surface

$F_{friction} = \mu N$

Coefficient of friction – μ

Downward force (normal force - N)

FRICTION LAWS

Two laws identified by **Guillaume Amontons**, and a **third** proved by **Charles-Augustin de Coulomb**, describe the **behaviour of dry friction**:

1. **Frictional force** is **proportional** to the **applied load** at the **boundary between the surfaces** (i.e. the **proportion of the object's weight pressing down into the surface**).
2. **Frictional force between objects** is **independent** of the **areas in contact**.
3. **Frictional force** is also **independent** of the **velocity at which two surfaces slide past each other**.

NEWTON'S LAWS OF MOTION

Isaac Newton's famous three laws of motion describe the behaviour of objects in ideal situations, and provide the key to understanding the link between force, movement, and momentum.

FUNDAMENTAL LAWS

First law:
An **object will remain at rest** or in a state of **constant uniform motion** unless **acted upon by an external force**.

Second law:
When a **force is applied to an object**, it experiences a **change of momentum** at a **rate directly proportional to the force**, and in the **same direction**.

Third law:
For every **action** there is an **equal but opposite reaction**. In other words, the **force that one body exerts on another** is **precisely balanced by a force exerted by the second body upon the first**.

FORCE, MASS, AND ACCELERATION

In order to **model** the **effects of forces**, **Newton** developed the **mathematical techniques** known as **calculus**, which deal with **rates of change** and the **values of various properties** at **specific moments** in a **changing system**.

In the **modern calculus notation (developed independently** by **Newton's rival Gottfried Leibniz)** the **second law** states that

$$F = \mathrm{d}p/\mathrm{d}t.$$

In other words, the **force** F is **equal** to the **rate of change of momentum** p, with respect to **time** t.

Because **momentum = mass × velocity**, the equation can be **rewritten** as

$$F = \mathrm{d}mv/\mathrm{d}t.$$

And because the **object's mass does not change**, this is also **equivalent** to

$$F = m \, \mathrm{d}v/\mathrm{d}t.$$

Here, $\mathrm{d}v/\mathrm{d}t$ is simply acceleration, so this is equivalent to the simple equation

$$F = ma.$$

In other words, force = mass × acceleration, and conversely acceleration = force/mass.

SUVAT EQUATIONS

This group of simple equations, taught in every classroom, allows us to calculate various aspects of mechanical motion in detail.

s = distance travelled
u = initial velocity
v = final velocity
a = acceleration
t = time

$$v = u + at$$

Final velocity = initial velocity + (acceleration × time).

$$s = ut + \tfrac{1}{2}at^2$$

Distance travelled = (initial velocity × time) + (½ acceleration × time squared).

$$s = \tfrac{1}{2}(u + v)t$$

Distance travelled = average velocity × time.

$$v^2 = u^2 + 2as$$

Final velocity squared = initial velocity squared + (2 × acceleration × distance travelled).

$$s = vt - \tfrac{1}{2}at^2$$

Distance travelled = (final velocity × time) - (½ acceleration × time squared).

GRAPHICAL TRICKS

Plotting motion on **graphs** is a **useful way** of **calculating various properties**:

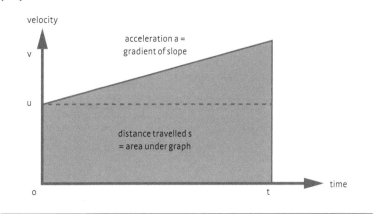

DIRECTIONAL MOTION

When **assessing** the **mechanics** of **moving bodies**, it's often useful to consider the **components of motion** and **force** in a **particular direction**. These can be **calculated using** some basic **trigonometry**.

If v is the **object's velocity** at a **particular angle** θ to an **arbitrary direction of interest**, then the **component of v in that direction** is **given by $v \cos θ$**, and the **component** in the **perpendicular direction** is given by **$v \sin θ$**.

The **same principles** can be used to **calculate** the **components** of **any other property of interest** (**force, acceleration**, etc.) **acting at a certain angle**.

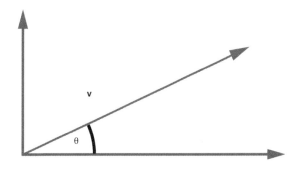

WORK, ENERGY, AND POWER

The three related concepts of work, energy, and power have their origins in elementary mechanics, even though today they are used in many other situations.

WORK = FORCE IN MOTION

Physicists measure the **way in which a force is "used up"** in terms of **work**. A **force does work** when it **moves a mass *m*** in a **particular direction** by an **amount *d*:**

$$\text{work done} = \text{force} \times \text{distance moved}$$

Work is measured in units called **newton metres**, known more commonly as **joules** after **Victorian scientist James Joule**.

Example: If a **force of 5 newtons** is **applied** to **move a mass 3 metres**, then **15 joules of work has been done**.

ENERGY = CAPACITY FOR WORK

Energy takes many forms but as a simple concept in mechanics, it is the **capacity of a system to do work**, and is therefore also **measured in joules**.

Like **momentum**, the **total energy** in a **closed system** is always **conserved** – it can be **moved from one form to another** but **cannot be created or destroyed**.

If a system contains 15 joules of **usable energy**, it should be possible to find a way of making it do 15 joules of **work**. However, **not all energy in a system *is* usable**, and an entire field of physics, **thermodynamics**, is devoted to describing the **function and transfer of energy**.

POWER = RATE OF WORK

Power is defined as the **rate at which work is done** or **energy is used**. It is measured in **watts**, where

$$1 \text{ watt} = 1 \text{ joule per second.}$$

From the simple example above, if a system has a **power of 15 watts**, it can exert **5 newtons** to **move the mass over 3 metres in 1 second**. If it has a **power of only 5 watts**, it will take **3 seconds to do the same work**.

MECHANICAL ADVANTAGE

Although it's impossible to get something for nothing in physics, a variety of devices called simple machines offer ways of making various mechanical tasks more manageable when power is limited.

DEFINING ADVANTAGE

Most of the devices known as **simple machines** offer ways of **reducing** the **force required** to **move** a **particular object** or **load**. We've already seen that

$$work = force \times distance\ moved$$

and understood that this means that if it takes a force of 5 newtons to move a mass over 3 metres, then 15 joules of work have been done.

But it's **also possible to do 15 joules of work** by **applying a force of 2.5 newtons over 6 metres**. The force is **halved**, even though the **distance** through which it must be applied is **doubled**.

Mechanical advantage is defined as the **ratio** between the **output force** exerted on an object or load F_{out}, and the **input force** applied, F_{in}:

$$MA = \frac{F_{out}}{F_{in}}$$

SIX SIMPLE MACHINES

Engineers going back at least as far as **Ancient Greece** based their ideas around six simple machines that offer mechanical advantage:

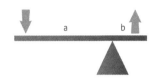

- **Lever**
  MA = a/b

- **Inclined plane**
  MA = length/rise

- **Wheel and axle**
  MA = radius $_{wheel}$ / radius $_{axle}$

- **Pulley**
  MA = n (the number of sections of rope supporting the load)

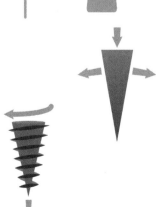

- **Wedge**
  MA = length/width

- **Screw**
  MA = 2pr/l
  (where l is the lead – the axial distance the screw travels during a full rotation)

POTENTIAL AND KINETIC ENERGY

The two most important forms of energy in simple mechanical systems are potential and kinetic: there are many familiar machines from everyday life that rely on transferring energy between these two forms.

POTENTIAL ENERGY

An object's potential energy is the **energy it has as a consequence of its position in relation to a force field**. This energy **doesn't really do anything useful**, but it has the **potential** to be **changed into other forms**.

Mechanics is mostly concerned with *gravitational potential energy*: energy resulting from an object's **location in Earth's gravity field**. In practice, this means its relation to an **arbitrary base level**, such as the **ground**, a **table**, or the **lowest point on a rollercoaster loop**. The **further the object has to fall**, the **more potential energy** it contains:

$$P.E. = mgh$$

where m is the object's **mass**, g is the **acceleration due to gravity**, and h is **height above the base level**.

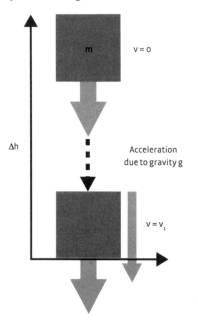

KINETIC ENERGY

Kinetic energy is the energy an object has due to its **motion**. Although **related to momentum**, it's described by a more complex equation:

$$K.E. = \tfrac{1}{2}mv^2$$

where v is the **velocity** of the moving mass.

ROLLERCOASTER PHYSICS

The **rollercoaster** is a classic example of the **relationship between kinetic and potential energy**. At the **highest point**, the cars' potential energy is **greatest** but they are **barely moving**. During the **drop**, potential energy is rapidly transformed to **kinetic energy** as the cars **pick up speed**. Kinetic energy is greatest at the **bottom of the dip**, and this **carries the cars back up** the other side of the slope, **regaining potential energy** as they **slow down**.

TYPES OF COLLISION

While the overall momentum of bodies involved in an interaction or collision is always conserved, that's not always the case for their kinetic energy. This helps to define two different types of interaction between objects.

ELASTIC COLLISIONS

Collisions or interactions in which the **overall kinetic energy in a system is conserved** are termed **elastic**. This means that the **sum** of the **kinetic energies** of **all particles in the system** remains the **same before and after their interaction**.

If objects with masses m_1 and m_2 and initial velocities u_1 and u_2 are involved in an elastic collision, then

$$\tfrac{1}{2}\,m_1 u_1^2 + \tfrac{1}{2}\,m_2 u_2^2 = \tfrac{1}{2}\,m_1 v_1^2 + \tfrac{1}{2}\,m_2 v_2^2$$

where v_1 and v_2 are the final velocities.

This means that for an elastic collision:

$$v_1 = \frac{m_1 \text{-} m_2}{m_1 + m_2}\,u_1 + \frac{2m_2}{m_1 + m_2}\,u_2$$

and

$$v_2 = \frac{m_2 \text{-} m_1}{m_1 + m_2}\,u_1 + \frac{2m_1}{m_1 + m_2}\,u_1$$

Elastic collisions are **very rare** in the **everyday world** – energy is **always transferred to or from other forms** due to the **internal movement of chemical bonds** – even in objects as small as **gas molecules**. Only collisions between **atoms** are **truly elastic**.

However, there are many systems that can be **treated as elastic** – either because the **losses and gains balance out statistically** (as with the **kinetic theory of gases**) or because they are **negligible on the scale being considered** (for instance in **collisions between pool balls**).

INELASTIC COLLISIONS

Far more widespread than elastic collisions, **inelastic collisions *usually* involve the loss of kinetic energy** to other forms, typically **heat** and **potential energy**.

Velocities of particles after an inelastic collision are given by:

$$v_1 = C_R \frac{m_2(u_2 \text{-} u_1) + m_1 u_1 + m_2 u_2}{m_1 + m_2}$$

and

$$v_2 = C_R \frac{m_1(u_1 \text{-} u_2) + m_1 u_1 + m_2 u_2}{m_1 + m_2}$$

where C_R is the system's **coefficient of restitution**, a measure of the collision's elasticity ranging from 0 (perfectly inelastic) to 1 (perfectly elastic).

A **perfectly inelastic collision** is one in which the **maximum possible amount of kinetic energy is lost** – typically when the **colliding objects stick together**.

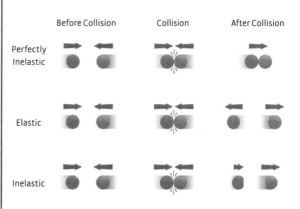

GRAVITATION AND ORBITS

Gravitation is the most familiar force in the Universe, and also the most commonly overlooked.
It is a force exerted by all objects with mass, pulling other objects towards them.

MECHANICS

UNIVERSAL GRAVITATION

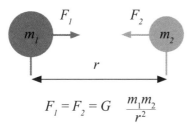

$$F_1 = F_2 = G \; \frac{m_1 m_2}{r^2}$$

Newton calculated that the **gravitational force** exerted between **two objects** is **proportional** to **both of their masses**, and **inversely proportional** to the **square of the distance between them**.

In other words, **doubling either mass doubles** the **attractive force between them**, but **doubling the distance reduces** the force to a **quarter of its former value**.

Note that both objects feel **equal and opposite forces of attraction (Newton's third law)**.

EXPLAINING ORBITS

Newton's **discovery of gravitation** was driven by an interest in explaining **Kepler's laws of planetary motion** (see p.20). By **modelling orbits** as **ellipses** on which, at any point, a **planet's tendency** to **keep moving in a straight line** is **precisely balanced by the pull of the Sun's gravity**, he was successfully able to link the **inverse square strength of gravity** with Kepler's own **distance/orbital period relationship**.

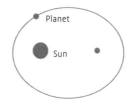

GRAVITY ON EARTH

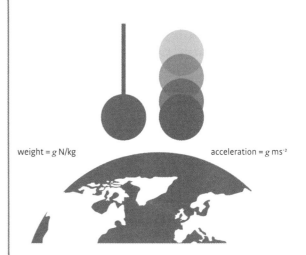

weight = g N/kg   acceleration = g ms^{-2}

At **Earth's surface**, the gravitational force is referred to simply as **gravity**. It is felt as a **downward force acting on all objects in Earth's gravitational field**. The force per kilogram is given by

$$F/m_2 = G \, m_1/r^2,$$

which works out at 9.81 newtons/kg.

Provided the **mass** of an object at or close to Earth's surface is **negligible** compared to that of Earth itself, then it will **always experience** the **same downward acceleration due to gravity**, denoted g. This is because, according to **Newton's second law**, force = mass × acceleration. Therefore:

$$F = m_2 \, a = \frac{G m_1 m_2}{r^2} \quad \text{and} \quad a = G \, m_1/r^2.$$

Unsurprisingly, therefore, g has a value of 9.81 ms^{-2}.

LAGRANGIAN MECHANICS

While Newton's laws neatly describe perfect systems of isolated particles, the reality of most physical systems is very different. Lagrangian mechanics offers a variety of tools that can be applied to real situations.

LIMITS TO NEWTON

Newton's laws are ideal for **calculating the orbits of planets**, the **ballistics of flying cannonballs** and the **movement of balls on a pool table**. But what about situations where **motion is constrained** – for instance, when a **rollercoaster** is anchored to a track that creates a **force in different directions from moment to moment**?

In 1788, French mathematician **Joseph-Louis Lagrange** proposed two new mathematical approaches that allow **Newton's laws** to be **used in a wider variety of situations**.

EQUATIONS OF THE SECOND KIND

The equations of the **second kind** take a more sophisticated approach that is **mathematically more complex**, but **does away with the need to consider the constraints separately** by changing the **configuration space** – the system of **coordinates** describing the **motion** – to whatever is most appropriate.

By stepping away from the traditional **three-dimensional** (x, y, z) coordinates of **Descartes**, the equations are able to consider an object's **properties at different points in a constrained system** while **ignoring the constraints themselves**.

EQUATIONS OF THE FIRST KIND

Lagrange's equations describe the changing dynamics of the *Lagrangian function* L. In most systems:

$$L = T\text{-}V$$

where:
T = *total* kinetic energy (sum of energies of all particles in the system)
V = total potential energy.

The equations of the **first kind** show how each **constraint** on a system can be described in terms of a mathematical equation (a **function**) and then applied to the overall description of how the Lagrangian **changes with time and position** in the system.

HARMONIC MOTION

Periodic or harmonic patterns of motion occur widely both in nature and in machines. An idealized form of this phenomenon, which is surprisingly common in the real world, is called simple harmonic motion (SHM).

REQUIREMENTS OF SHM

Simple harmonic motion was discovered in 1732, when **Daniel Bernoulli** applied Newton's laws to points along a **vibrating string**:

- The **force acting on an object** grows **larger** with **displacement** from a **notional point of rest**.
- This **force always acts** in the **opposite direction** to the **displacement** (i.e. to **restore the string to the centre**).

Harmonic motion occurs in situations like this because of a **repeated exchange** of **kinetic** and **potential energy** – the string **comes to a halt** at its **maximum displacement** (when its **potential energy** is **greatest**), and has **peak kinetic energy** (and **velocity**) as it **crosses the centre line**, which causes it to **overshoot** to the other side.

MODELLING SHM

One of the neatest aspects of SHM is that it can be **modelled** in the form of **waves**. The amount of **displacement**, the **velocity** of the **displaced object** and the **strength** of the **restoring force** all take the form of **mathematically perfect sine waves**, offset to each other:

Where t = time, T is the period of oscillation, and ω is the angular frequency, measured in *radians* per second – calculated from the more familiar frequency f (measured in cycles per second) as $2\varpi f$ (or $2\pi/T$).

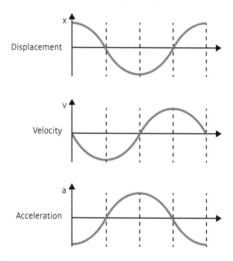

SHM crops up in some unexpected places, such as alternating electric currents (see p.101).

SHM EXAMPLES

Pendulums and other swinging weights.

Oscillating weights on the end of springs.

Musical instruments and the sound waves they produce.

A hanging beam twisting in one direction and then the other.

In addition, SHM crops up in some unexpected places, such as alternating electric currents (see p.101).

ANGULAR MOTION AND RIGID BODIES

Harmonic motion crops up in some surprising places – and is particularly important for modelling circular motion and uniform rotation. This in turn leads to ways of analysing more complex forms of motion.

ROTATING BODIES

In any **rotating system**, be it a **spinning solid body** or an **object circling a central axis**, the behaviour of a **rotating point** can be considered as **simple harmonic motion** in **two separate dimensions** – viewed from **above**, the point **moves up and down** and **back and forth** in **two interlocking cycles** that are both described by **sine waves**.

Just as **Newton's second law of motion** states (in one form) that

$$F = m \, dv/dt$$

for **linear motion**, so this has a **direct parallel** in **Euler's equation** of **rotational motion**:

$$\tau = I \, d\omega/dt.$$

In this case, τ (the Greek letter tau) is the **rotation force** acting on the object, known as **torque**, I is the object's **moment of inertia** (**resistance to turning**) and $d\omega/dt$ is the **rate of change in angular velocity** ω (in other words, its "**rotational acceleration**"). As with other forms of harmonic motion, ω is measured in *radians* per second.

EULER'S DISCOVERY

Swiss mathematician **Leonhard Euler** discovered his **equation for rotational motion** while studying the **physics of ships** and their **interactions with waves**. He realized around 1736 that seemingly **chaotic cycles** of **movement**, **rolling**, and **pitching** can be broken down into two elements: **translation (movement through space, described by Newton's second law)** and **rotation (movement around an axis, described by Euler's equation)**. The same principle can be applied to the **motions of any rigid body**, with **applications far beyond the behaviour of ships**.

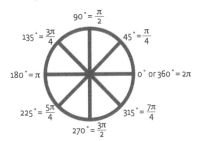

WHAT ARE RADIANS?

A **radian** is an alternative **unit of angular measurement** to the more familiar degrees. Just as there are 360 degrees in a circle, there are also 2ϖ (roughly 6.2831853) radians in a **circle** (and also in a **complete cycle of harmonic motion**). A radian is equivalent to $180/\varpi$ degrees, or roughly 57.296°.

Radians seem confusing, but they make some calculations a lot easier. Consider that the **circumference of a circle** is $2\varpi r$ (where r is the **radius**): a distance of r measured along the circumference will "**subtend**" an angle of 1 radian at its centre.

General planar motion = Translation + Rotation

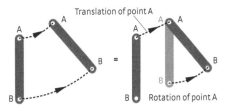

39

ANGULAR MOMENTUM

Just as objects moving through space have momentum due to their speed and mass, so rotating bodies have their own equivalent: angular momentum.

CALCULATING ANGULAR MOMENTUM

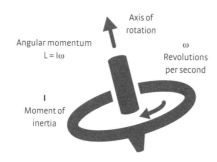

Axis of rotation

Angular momentum
L = Iω

ω
Revolutions per second

I
Moment of inertia

The momentum of rotating objects has both a **magnitude** (**strength**) and an **instantaneous direction** – albeit one that is **constantly changing**. It is calculated from the simple formula:

$$L = I\omega$$

where L is the **angular momentum about an axis of rotation**, I the **object's moment of inertia** (its **resistance to rotation**), and ω its **angular velocity** in radians per second.

This is directly comparable to the equation for **familiar linear momentum** p:

$$p = mv$$

in which m is the object's **mass** (also a **measure of inertia**) and v its **linear velocity**.

CONSERVATION OF ANGULAR MOMENTUM

Just as with normal momentum, the angular momentum of an **object** or **enclosed system** is **conserved** – it remains the **same** unless **influenced by outside forces**. This has important effects because, while the **mass of objects usually remains constant without external influence**, **moment of inertia** is a **more complex property** that **depends not just on mass**, but also on its **distribution in relation to the axis of rotation**:

- If mass is **spread far from** the **axis of rotation**, the moment of inertia is **greater**.
- If the **same mass** is **concentrated close** to the axis, the moment of inertial is **less**.

Therefore, in order to conserve momentum:

- If the mass in a system becomes more **widely spread**, its **angular rotation must slow down**.
- If the mass is **concentrated**, its **angular rotation must speed up**.

EXAMPLES

- Pirouetting **ice skaters** extending or drawing in their arms to control their moment of inertia and angular velocity.
- The rapid rotation of **pulsars** – superdense objects formed when massive stars collapse into city-sized balls.
- The increasingly rapid rotation of **newborn stars** as they draw in matter from surrounding gas clouds.

FICTITIOUS FORCES

Fictitious forces are ones that only seem to affect objects because of our frame of reference. They do not really exist except as the net effect of interactions between various real forces.

CENTRIFUGAL FORCE

Centrifugal (centre-fleeing) force is the effect that **seems to tug a ball on the end of a string away from you** when you swing it around your head.

It arises from **Newton's first law**, since at any point on its arc the ball will have a tendency to **fly off in a straight line unless another force acts on it**.

The **true force** acting on the ball in this situation is a *centripetal* (**centre-seeking**) one – the **tension of the string** pulling it towards your hand.

At **each point along its arc**, this force **counters** the **ball's tendency to move off from** a **circular path**. The result is a **curved path**.

Orbits involve a very similar effect – though in the case of an **orbiting satellite** it's **Earth's gravity** that provides the **centripetal force**.

THE CORIOLIS EFFECT

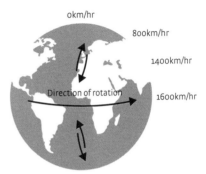

okm/hr
800km/hr
1400km/hr
Direction of rotation
1600km/hr

Coriolis "forces" are an effect created when an **observer on the surface of a solid rotating body** such as a **planet** observes the **motion of unattached, free-moving objects**.

Objects moving in a **clockwise-rotating frame of reference** appear to experience a **force to their left**, while those in an **anticlockwise** frame of reference appear to be **pushed to their right**.

In both cases the objects **appear to follow curving paths**, with the effect getting **stronger further from the centre of rotation**.

Because the frame of reference is **anticlockwise in the northern hemisphere** and **clockwise** in the **southern**, their **coriolis "forces" act in different directions**.

Because it **varies with latitude**, the effect also causes **large bodies of inward- or outward-flowing air or water** to follow **spiral patterns of rotation**.

DOWN THE PLUGHOLE!
Coriolis forces **only create rotation on very large scales** – sadly the claim that they make **water in opposite hemispheres drain down plugholes in different directions** is just an **urban myth**.

CHAOS

The term "chaos" has a very specific meaning for physicists and mathematicians – it refers to systems within which small, apparently negligible changes can ultimately give rise to large and unpredictable results.

DEFINING CHAOS

In everyday usage, chaos implies something that follows **no rules** and is entirely **unpredictable**. In physics, **chaotic systems can be governed by rules** that are simple and **entirely understood**.

The problem is "**sensitivity to initial conditions**". **Complex relationships** between objects in a system and **incomplete knowledge** of their properties make their **future evolution impossible to predict beyond a certain stage**.

POTENTIAL FOR CHAOS?

When **simulating the real world**, we can never hope to begin with **perfect** and **exhaustive knowledge** of the **position** and **motion** of **every component in a system**. But we can identify situations with **chaotic potential**:

$$x^2 + y^3$$

- Systems governed by **non-linear equations** involving **higher powers of certain variables** are **most prone to chaos**. **Multiplying the variable with itself** makes matters worse.

$$(2x + y) / z$$

- Systems with **linear equations** (where variables are simply multiplied and divided by other values, and **added** or **subtracted**) are **less likely to be chaotic**. Small changes or **inaccuracies in variables** here lead to only **small differences in outcomes**.

THE THREE-BODY PROBLEM

Known since the eighteenth century: it's **easy to predict a single planet's orbit around the Sun** into the **far future**, but **add a second planet** and things grow far **more complex** as the **gravitational pulls** of the **three bodies interact**.

THE BUTTERFLY EFFECT

Outlined by meteorologist **Edward Lorenz** in 1972. While running early **computer models** of the **weather**, he found that taking the **shortcut** of "**rounding down**" a **key variable** sent his **simulations** in a completely **different direction**.

The flap of a butterfly's wings in the Amazon can give rise to a tornado in Texas.

THEORIES OF MATTER

Philosophers and scientists have been trying to understand the nature of matter for at least three thousand years. Our current atomic theory allows us to predict how different materials will behave in different situations.

WHAT IS MATTER?

Matter is simply the physical stuff that makes up the Universe. Many **early philosophers** made an important **distinction between the physical and mental or ideal realms** – for example, **Plato** imagined the objects of the everyday world to be mere **shadows of ideals existing in a higher plane**.

ANCIENT ELEMENTS

One popular theory in **classical Greece** saw all matter as a mix of **four elements: earth**, **air**, **fire**, and **water**. The precise blend in a particular substance could explain its properties.

EARLY ATOMISTS

Early Greek atomists such as **Democritus** (c. 400 BC) imagined a Universe made up of **tiny indivisible atoms** separated by **empty space** (void).

$$C + O_2 \rightarrow CO_2$$

ATOMIC THEORY

From the late eighteenth century, **John Dalton** and others revived the **atomic theory** to explain why **chemical reactions** often seemed to involve **combining simple ratios** of materials.

ELEMENTS IN ORDER

In 1869, **Dmitri Mendeleev**'s first **periodic table** established patterns among the **known elements**, helping to **predict missing ones** and raising questions concerning why these **patterns** arose.

ATOMIC STRUCTURE

In 1897, **J. J. Thomson** discovered **electrons**, the **first subatomic particles**. These led to the identification of a **distinct atomic nucleus** in 1911, and the **Bohr model** of the atom (with electrons following **distinct orbits** around the **nucleus**) in 1913.

QUANTUM PHYSICS

Breakthroughs in the 1920s revealed that on the **smallest scales**, subatomic particles have **wavelike properties** and can **behave in unpredictable ways**.

BROWNIAN MOTION

The most direct **proof of atomic theory** came from a phenomenon called **Brownian motion** – the jiggling movement or "**random walk**" of tiny particles such as pollen grains when they are **immersed in water**. Botanist **Robert Brown** reported the effect in 1827.

PROOF OF ATOMISM

In 1905, **Albert Einstein** showed how Brownian motion could be explained if the pollen grains were being pushed around by **collisions with invisible water molecules**.

STATES OF MATTER

In most everyday situations, matter is found in one of three different states – solid, liquid, and gas. But there's also a fourth state, known as a plasma, that can arise in some extreme conditions.

TYPES OF MATTER

SOLIDS
A material whose **atoms are strongly bound together**, allowing it to **maintain its shape** without a container.

LIQUIDS
Material with **weaker attractive forces between atoms or molecules**, which are **overcome by gravity**, causing it to **flow** and fill the **lowest level** in a container.

GASES
Materials in which **atoms or molecules are separated entirely** from each other and **move round at speed, filling their container**.

PLASMAS
Gas-like fluid in which the **particles lose their electrons** to become **electrically charged ions**. In general, **particles in solids have the least energy** while **those in plasmas have the most**.

CHANGING STATE

Although we're used to materials changing from **solid** to **liquid** to **gas** (and vice versa), **all three states can actually transform from one to another**:

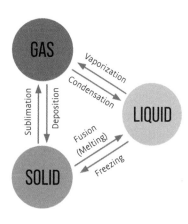

LATENT HEAT

Changes of state involve breaking or forming **bonds between atoms**. Perhaps surprisingly, **breaking bonds absorbs energy** while **forming them releases it**. The amount of energy released or absorbed during these processes is known as the "**latent heat**" of **fusion** (for the **solid → liquid** transition) or **vaporization** (for **liquid → gas**).

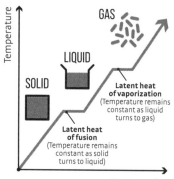

PHASE DIAGRAMS

The **precise state of matter** that a particular substance **takes on** is determined by its own **chemistry**, but also by **conditions around it** – specifically **temperature** and **pressure**. The relationship between state, temperature and pressure is shown by a **phase diagram**:

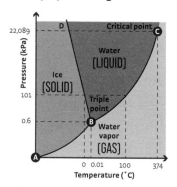

- **Triple point**: Conditions in which solid, liquid, and vapour can **coexist**.
- **Critical point**: Where phase boundaries begin to **dissolve**.

SOLIDS AND THEIR STRUCTURES

Solid materials involve atoms or molecules that are tightly bound together to form groups large enough to mostly prevent the passage of other materials through them. Solids may be either crystalline or amorphous (shapeless).

CRYSTAL LATTICES

Crystalline lattices can take on various forms, depending on the **exact way that atoms or molecules bond together**. These are:

CUBIC
All three axes perpendicular and of equal length.

TETRAGONAL
Three perpendicular axes, two with equal length.

ORTHORHOMBIC
Three perpendicular axes, all with different lengths.

HEXAGONAL
Three axes of equal length separated by equal angles, with a fourth axis perpendicular to the plane of the first three.

MONOCLINIC
Two perpendicular axes, all three axes of equal length.

TRICLINIC
No perpendicular axes and all axes are of different lengths.

RHOMBOHEDRAL
No perpendicular axes, but all of equal length.

ALLOTROPES

Allotropes are **different forms** that **a substance can take**. Although most substances can exist as solid, liquid, or gas, the term allotrope **usually applies to different solid forms**:

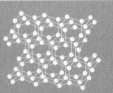

Diamond:
Cubic crystal

Graphite:
Hexagonal crystal (planar)

Buckminsterfullerene:
Hexagonal crystal (ball-shaped)

AMORPHOUS SOLIDS

Not all solids have atoms in a **repeating crystalline structure** – many have **less orderly arrangement** with **no parallel rows or planes of atoms**. These amorphous solids include **glass** and many **polymers**, including **plastics**.

DEFORMATION AND ELASTICITY

Most solids can alter their shape to at least some extent without completely breaking apart. Bonds between molecules can be stretched or rearranged in ways that affect a material's large-scale or "bulk" properties.

STRESS AND STRAIN

Stress is a **measure of the force that acts through a solid** due to **each particle experiencing push or pull from its neighbours**.

Stress = **deforming force**/square metre of material cross-section

Stress is measured in **pascals**, the same units used for **pressure**:

$$1 \text{ Pa} = 1 \text{ N/m}^2$$

1 pascal = 1 newton per square metre

Strain measures the **deformation of a material** in **response to stress**. It's measured as the **change of length in a particular axis relative to the material's overall length**, and is a **simple number**:

Strain = **change in length** per **metre of material length** = $\Delta L/L$

DUCTILITY AND BRITTLENESS

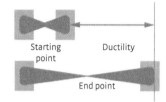

Starting point — Ductility — End point

Materials respond to **force** in different ways. In some **metals** (for example, **copper**), the **atoms easily slip past each other and rearrange**, allowing them to be **stretched out into long strands** – they are said to be **ductile**.

Other materials are **brittle** – they **resist stress** up to a point with **very little deformation**, then **undergo catastrophic failure** in which their **internal bonds are completely broken**.

HOOKE'S LAW

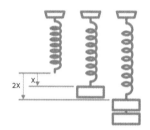

The principle that the **amount of deformation created in a given elastic material is proportional to the force acting on it** was discovered by **Robert Hooke** in 1676.

$$F = kx$$

(where k is the material's **"spring constant"**)

Hooke wrote his law down in a **Latin anagram**:

Ceiiinosssttuu

which unravels as:
Ut tensio, sic vis.

Translated, this means "As the extension, so the force."

YOUNG'S MODULUS

Also known as the **modulus of elasticity**, Young's modulus measures the **relationship between stress and strain** in a material:

$$E = \text{stress/strain}$$

As with stress, Young's modulus is **measured** simply in **units of pressure**. However, it's typically a **large quantity** and is therefore often measured in **gigapascals** (GPa = billions of pascals).

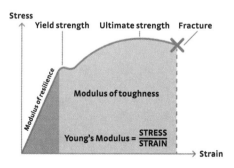

46

FLUID MECHANICS

Fluids are materials bound together by relatively loose bonds that are constantly being broken and reformed. This allows a fluid to flow under the influence of gravity and other forces and fill its container, without fragmenting entirely.

ARCHIMEDES' PRINCIPLE

When an object is placed in a fluid, it is supported by an **upward force (buoyancy) equivalent to the weight of fluid it displaces**. The object appears to have lost weight and either **floats**, **hangs**, or **sinks more slowly** than it would fall in air.

EUREKA!

Archimedes supposedly discovered his principle when asked to assess the **purity of a gold crown**. However, in reality the principle wasn't needed to solve the problem – he simply used the **amount of water displaced** by the fully submerged crown to **measure its precise volume**, and could then easily work out its **density**.

FLUIDS IN WEIGHTLESSNESS

In the **weightless** conditions of **space**, fluids pull themselves into **spherical shapes** thanks to surface tension – the spherical shape ensures that **all the bonds across the surface are of equal length**.

BUBBLES

Adding **soap** to **water** weakens its **intermolecular bonds** so that **bubbles** can be formed and can survive for some time without collapsing.

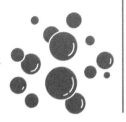

SURFACE TENSION

In the middle of a liquid, **bonds between molecules are in balance, pulling them in all directions at once**. At the **surface**, bonds only pull molecules **inwards** and **across the surface**. This effect, known as **surface tension, resists the breaking of the surface**.

Surface tension **prevents different liquids from mixing**, and **allows surfaces to support lightweight objects** such as insects.

Surface tension is highest in liquids, like **water**, with particularly **strong intermolecular bonds**.

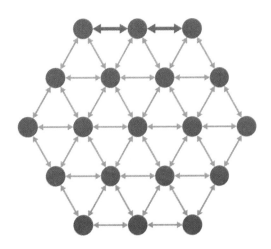

FLUID DYNAMICS

Fluid dynamics is the branch of physics that deals with flows of fluid, including both liquids and gases. Rather than breaking them down into individual particles, it treats fluids as continuous substances in which a disturbance of one region affects all the others.

KEY CONCEPTS

- **Incompressibility**: Fluid dynamics assumes that the fluids it studies **cannot be compressed by pressure** – something that's **generally true for liquids but not for gases**.

- **Ideal fluids**: These model fluids have **no viscosity** (internal friction) to prevent them from **deforming**.

- **Shear stress**: The **external force** applied per unit area of fluid material ***parallel*** to the force.

- **Shear strain**: The change in the fluid's **dimensions** in response to the **shear stress, compared to its overall size**.

TAKING FLIGHT

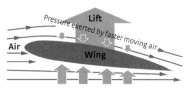

Pressure exerted by slower moving air

Bernoulli's effect allows **aircraft** to **fly** – **aerodynamic design** creates **slower air flow beneath the wing** and **faster air flow across its upper surface**. The greater **air pressure** beneath the wing lifts the aircraft up.

NEWTONIAN VS NON-NEWTONIAN FLUIDS

Newtonian fluids are "normal" ones in which **stress is proportional to the rate of change of strain at every point**, and **viscosity** is therefore **constant**.

Non-Newtonian fluids have **variable viscosity** that can be affected by **factors such as time** and the **stress** already experienced. These unpredictable fluids include **ketchup** and **paint**.

VISCOSITY

The concept of viscosity is somewhat similar to **Young's modulus** (a measure of the **elasticity of solid materials** – see p.46), but because fluids **deform rapidly** it's usually calculated as:

- Viscous substances
- Water (viscosity = 1 millipascal second)
- Non-viscous substances

Viscosity = shear stress/rate of change in shear strain

(measured in pascal seconds)

BERNOULLI'S PRINCIPLE

In 1738, Swiss mathematician **Daniel Bernoulli** made a counterintuitive discovery: **speed of fluid flow is inversely proportional to the pressure exerted by the fluid**.

In other words, **when a fluid increases its speed** (for example, when the pipe it is passing through narrows) **it exerts less pressure**.

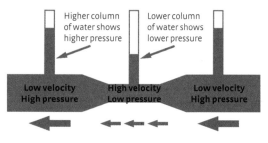

Higher column of water shows higher pressure

Lower column of water shows lower pressure

Low velocity High pressure

High velocity Low pressure

Low velocity High pressure

IDEAL GASES

An ideal gas is a material in which atoms or molecules are entirely independent of each other, floating around with no significant forces to draw them together. In these circumstances, a gas will spread out to fill the volume of its container.

GAS LAWS

Three simple laws describe the **behaviour** of a fixed amount of gas in a container when various conditions change:

BOYLE'S LAW

The **pressure** exerted by the gas is **inversely proportional to the volume** it contains, **provided the temperature remains the same**.

Squeezing the gas into a smaller volume **increases** its pressure; allowing it to **expand decreases** the pressure.

P is **inversely proportional** to V at constant T.

CHARLES'S LAW

The **volume** occupied by a gas is **directly proportional to its temperature** if its **pressure remains the same**.

Heating the gas causes it to **expand**, if it is allowed to do so freely, while **cooling** causes it to **contract**.

V is proportional to T at constant P.

GAY-LUSSAC'S LAW

The **pressure** exerted by a gas is **directly proportional to its temperature** if its **volume remains the same**.

Heating the gas in a container of **fixed size** causes its **pressure to rise**, while **cooling** causes a pressure **drop**.

P is proportional to T at constant V.

AVOGADRO'S LAW

But what happens if we **change the amount of gas in a container**? Avogadro's law says that if **pressure and temperature are held constant**, the **volume occupied by a gas is proportional to the amount of gas present** (denoted by *n*).

V is proportional to *n* at constant T and P.

MEASURING GASES

Quantities of gas are measured in **moles** – the **number of atoms in the gram equivalent of** the **atomic or molecular mass**.

For example:
- One mole of helium (atomic mass = 2) weighs 2 g.
- One mole of molecular oxygen O_2 (atomic mass 16) weighs $2 \times 16 = 32$ g.

IDEAL GAS LAW

All of these gas laws can be combined into a **single equation** that describes the **behaviour of any ideal gas**:

$$PV = nRT$$

where R is a constant, known as the **ideal gas constant**.

One mole **always contains the same number of atoms** ...

Avogadro's Number

6.022 x 10²³

... and occupies the same volume of space at "standard temperature and pressure" (0 °C and one bar):

Molar volume =

22.4 LITRES

KINETIC THEORY OF GASES

The kinetic theory explains and models the behaviour of gases in deceptively simple terms by considering the way that each particle in a gas will behave as conditions in the overall gas are changed.

KEY ASSUMPTIONS

An **ideal gas** is made of **individual atoms** or **molecules** that **only interact with each other through collisions**.

The **number of particles** is **so large** that their behaviour can be treated **statistically**.

Particles move at **speeds** dependent on the gas's *absolute* temperature. The **temperature** of the gas reflects the **kinetic** (motion) **energy of molecules** according to the equation $KE = \frac{1}{2} mv^2$.

The **pressure** a gas exerts is due to **collisions between the particles and the walls of their container**. Collisions between particles are *elastic* (the **total kinetic energy** shared between the particles **remains the same**), but those between particles and the container are *inelastic* (**kinetic energy is transferred** into other forms).

Gravitational pull on individual molecules is **negligible**.

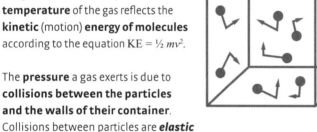

THE THEORY IN PRACTICE

Increasing the **temperature** of a gas in a **fixed container** increases the **speed of its molecules** and the **rate and strength of collisions with the container**, increasing the **pressure (Gay-Lussac's law)**.

If the container can **expand**, then **raising the temperature** and the **rate of collisions** will **push the walls outwards, increasing the volume** while **keeping the overall pressure steady (Charles's law)**.

Adding more gas without changing temperature increases the rate of collisions with the container. If the container can **expand**, it will do so until it **returns to its original pressure (Avogradro's law)**.

Increasing the size of the container **reduces the rate of collisions** and **lowers the pressure**. Conversely, **reducing the size increases the rate of collisions** and **raises the pressure (Boyle's law)**.

MAXWELL–BOLTZMANN DISTRIBUTION

Statistical treatment means that the **unknowable properties** of countless individual objects can be **modelled in terms of their distribution** around an **average**. The complex equations that describe this were worked out by **James Clerk Maxwell** and **Ludwig Boltzmann**.

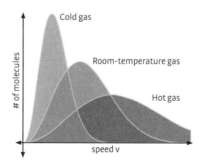

Cold gas

Room-temperature gas

Hot gas

\# of molecules

speed v

STATISTICAL PHYSICS

The **kinetic theory of gases** was a huge breakthrough in the story of physics as a whole – it showed for the first time the **power of statistical models**, in which the **properties of individual particles** are **not so important as the statistical average of their behaviour**.

MATTER

CHEMICAL ELEMENTS

A chemical element is a substance whose atoms have unique properties different from the atoms of any other element. Ninety-four elements are known to exist in nature, while a further twenty-four have been manufactured artificially using nuclear reactors.

ELEMENTAL PROPERTIES

ATOMIC NUMBER

An element's **atomic number** represents the **number of protons** (**positively charged**, relatively heavy particles) in its **central nucleus**. Because atoms are only **stable** when they are **electrically neutral** (with **no overall charge**), the atomic number also indicates the **number of electrons** (negatively charged, lightweight particles) in **orbit around the nucleus of a lone atom**.

ATOMIC MASS

Most elements have an atomic mass measured in **multiples of a simple "atomic mass unit"** (a.m.u.). The **atomic mass unit** is defined as **one-twelfth of the mass of a single atom of carbon-12**.

 /12

A carbon-12 atom contains **six protons**, **six neutrons**, and **six electrons**.

ISOTOPES

Some elements can have atoms with **different masses**, known as **isotopes**. While each isotope has

the **same number of protons in its nucleus**, the **number of neutrons can vary**. Isotopes of the same element all have the **same chemistry**, so when the atomic mass of a pure sample is calculated, it **may not be a whole number**.

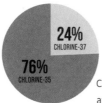

24% CHLORINE-37

76% CHLORINE-35

Chlorine = average atomic mass: 35.45

DISCOVERING THE ELEMENTS

Native elements – **metals** and **non-metals** such as **sulphur** that exist in **pure form in nature** – have been known since ancient times.

TIMELINE

Prehistory to 3000 BC Pure iron was discovered in meteorites that fell to Earth in prehistoric times, but the means of extracting it from mineral ores such as iron oxide was only discovered around 3000 BC

1669 The first new element, phosphorus, is discovered by Hennig Brand

1766 Hydrogen is identified as a distinct gas by Henry Cavendish

1777 Oxygen is identified as a new element by Antoine Lavoisier

1868 Helium is discovered by Janssen and Lockyer based on its fingerprint in the spectrum of sunlight

1875 Gallium is discovered after Dmitri Mendeleev's prediction based on a gap in his periodic table of elements

1898 Polonium is the first unstable element to be discovered, found by Marie and Pierre Curie as a relatively short-lived product of uranium

1937 Technetium is manufactured using an early particle accelerator – the first synthetic element to be made on Earth

2002 Oganesson, the heaviest element known, is synthesized for the first time in Russia

THE PERIODIC TABLE

The periodic table is a powerful tool for understanding the chemistry, physical properties, and internal structure of different elements.

DMITRI MENDELEEV

The periodic table is usually credited to Russian chemist **Dmitri Mendeleev**, who published an **early version** in 1869 and used it to **predict elements** (**gallium** and **germanium**) that had **not yet** been **discovered**.

GROUPING THE ELEMENTS

The periodic table arranges elements into **vertical groups according to their similar chemistry**, and **horizontal periods of increasing mass**. In addition, elements are often grouped into **several distinct blocks based on shared properties**.

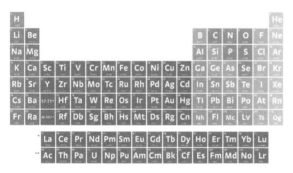

- **Hydrogen**: The **simplest element, highly reactive** with a **single electron in its outer shell**.
- **Alkali and alkaline earth metals**: These **shiny** metals have **unstable atomic structures** – they **react readily** (and **sometimes violently**) **with other elements to form compounds**.
- **Transition metals**: These **less reactive metals** can form a wide range of **compounds** through **different types of chemical reaction**.
- **Post-transition metals**: These **brittle metals** have **different chemical properties from** the **transition metals** and are **grouped separately**.
- **Metalloids**: These elements have **properties** between those of **metals and non-metals**.
- **Reactive non-metals**: This handful of elements have **low melting and boiling points** and **usually form** the "**other half**" of **compounds with metallic atoms**.
- **Noble gases**: These **non-metals** have **stable atomic structures** and therefore **do not undergo chemical reactions**.
- **Lanthanides and actinides**: These rows of **heavy metallic elements** are **traditionally separated** in order **to keep the dimensions of the periodic table manageable**.

ELEMENT DETAILS

Key:

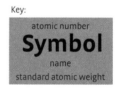

UNDERLYING STRUCTURE

The **pattern** of **rows** and **columns** in the periodic table reflects the way in which **subatomic electron particles fill up various shells within the atoms of each element**. The **number of elements** in each period **tells us something important about atomic structure** (see p.53).

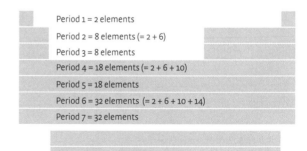

Period 1 = 2 elements
Period 2 = 8 elements (= 2 + 6)
Period 3 = 8 elements
Period 4 = 18 elements (= 2 + 6 + 10)
Period 5 = 18 elements
Period 6 = 32 elements (= 2 + 6 + 10 + 14)
Period 7 = 32 elements

INTRODUCING ATOMIC STRUCTURE

Although physicists have discovered a host of elementary particles and interactions, most of the chemical behaviour of elements can be explained using a simple model involving three particles.

SIMPLE ATOMS

  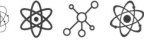

Atoms consist of **negatively charged lightweight electrons** orbiting around a **positively charged nucleus** that contains most of an atom's **mass**.

Because individual atoms are **electrically neutral**, the **charge on the nucleus** is **equal and opposite to the combined charge of the electrons**.

Nuclear particles called **protons** have an **equal but opposite charge to electrons**, so a **neutral atom contains as many protons as it does electrons**.

All but the simplest nuclei also contain **uncharged particles called neutrons**, each with **roughly the same mass as a proton**.

The **mix of particles** in an atom can therefore be worked out from its **atomic number** (position on the periodic table) and **relative atomic mass**.

The existence of **isotopes** with **different masses** complicates matters, so most descriptions of atoms specify **which isotope is being discussed**.

EXAMPLES

- **Simple hydrogen atom**: One proton orbited by one electron.

- **Deuterium (hydrogen-2)**: One proton and one neutron orbited by one electron.

- **Helium-4**: Two protons and two neutrons orbited by two electrons.

- **Carbon-12**: Six protons and six neutrons orbited by six electrons.

- **Carbon-14**: Six protons and eight neutrons orbited by six electrons.

ELECTRON ORBITALS

The patterns of the periodic table are caused by the way in which **electrons are "added" to the atom**. Each **horizontal period** corresponds to the **filling of a certain set of orbital shells that can each hold a pair of electrons**.

Orbital	Description	Electrons	Periods
s orbital	Spherical	2	1 to 7
p orbital	Three perpendicular double-lobed shells	$3 \times 2 = 6$	2 to 7
d orbital	Five double-lobed shells	$5 \times 2 = 10$	4 to 7
f orbital	Seven double-lobed shells	$7 \times 2 = 14$	6 and 7

Each period of the periodic table covers the **filling of a complete set of orbitals relevant to that period**, so the **filling sequence** goes: 1s, 2s, 2p, 3s ... etc.

ELECTRON CONFIGURATIONS

The arrangement of electrons in specific elements is described using **superscripts** to indicate **how many electrons are in a particular orbital**.

Element	Atomic number (= number of electrons)	Electron configuration
Hydrogen	1	$1s^1$
Helium	2	$1s^2$
Lithium	3	$[He]\ 2s^1$ *
Carbon	6	$[He]\ 2s^2\ 2p^2$
Neon	10	$[He]\ 2s^2\ 2p^6$
Silicon	14	$[Ne]\ 3s^2\ 3p^2$

* The symbol for the element with a complete set of shells at the right of the previous period is used to save writing out the complete configuration.

CHEMICAL BONDING

Chemical bonds combine atoms together in a variety of ways to form molecules. The nature and strength of the bond that is formed depends on the identity and relationship of the atoms involved.

SEARCH FOR STABILITY

The driving force in chemical bonding is the **urge for the atoms involved to reach a stable configuration with a complete outer shell of electrons**. This gives rise to **three key types** of chemical bond.

IONIC BONDS

An element that needs to **shed electrons for stability** donates them to one that needs to **gain them to complete its outer shell**. As a result, **two ions form** with **positive and negative charges**, which are then **bound together by electrostatic attraction**.

COVALENT BONDS

Two elements that both need to **gain electrons for stability** merge their outer electron shells and **share one or more electrons each to fill the gaps in the outer shell**.

METALLIC BONDS

Metallic elements with an **excess of electrons** in their outer shell **shed them** *en masse* to **form a crystalline structure** with **positively charged ions bound together by a "sea" of electrons scattered between them**.

CARBON-GROUP BONDS

Sitting in **Group 4 of the periodic table**, **carbon** and its **heavier relatives** all have **half-filled outer electron shells**. This makes them **relatively stable** but also means that they **can form up to four bonds with their neighbours**. This is why **carbon**, in particular, forms the **basis of a huge array of complex "organic" chemicals**.

CARBON BONDS

In **organic compounds**, carbon atoms can form **one, two, or three bonds with each other** while **still being able to bond together to other atoms**.

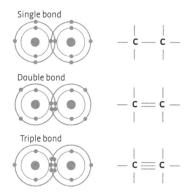

- Single bond = "-ane" suffix in name.
- Double bond = "-ene" suffix.
- Triple bond = "-yne" suffix.

THE BENZENE RING

In a handful of molecules, carbon forms **rings of six atoms** with a **single bond to another element or group**. The simplest example is **benzene**, C_6H_6. In this case, the carbon atoms each **donate an electron to a delocalized ring that is shared between them**. In effect, **single and double bonds alternate rapidly**.

CHEMICAL REACTIONS

*A chemical reaction is a process in which molecules of different materials are broken apart
and recombined, transforming a set of reactant chemicals into a set of products.*

REACTION EQUATIONS

Chemical reactions are usually described using
equations. For example, the simple reaction of **sodium**
with **chlorine** to give **sodium chloride** is written:

$$2Na \text{ (s)} + Cl_2 \text{ (g)} \rightarrow 2NaCl \text{ (s)}$$

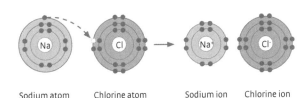

| Sodium atom | Chlorine atom | Sodium ion | Chlorine ion |

The total number of **atoms** of each **element** is the
same on **either side** of the arrow, and **all the elements
involved** are described in their **"correct" form**. So,
because **chlorine** is usually found as the **diatomic Cl_2
molecule**, **two atoms** of sodium are required to form
two molecules of NaCl.

- The arrow indicates a reaction's **preferred direction**.
  However, many reactions are **reversible**, and have a
  double-ended arrow in the middle: ↔. In some cases,
  the **reversibility requires special conditions**, but in
  other cases it **happens easily**, allowing the reaction to
  reach an **equilibrium point** as the **concentrations** of
  reactants and **products** vary.

The **bracketed letters** after each reactant or product
indicate its **physical state – solid, gaseous**, or **liquid –**
and are optional.

REACTIONS AND ENERGY

Breaking apart bonds within **molecules** requires
energy, while **forming new bonds releases energy**.
Many reactions therefore require the **addition of
external energy** (such as **heat**) to begin, but **release
energy** of their own **once underway**. Reactions
that **absorb more energy than they release** are
endothermic, while those which **release more
energy than they absorb** are **exothermic**. Those that
require no **additional energy** to **get them started** are
exergonic.

REDUCTION AND OXIDATION

Many reactions involve the **transfer of electrons**
from **one chemical species** (an **atom, molecule** or
electrically charged ion) **to another**. For historical
reasons, the terms **reduction** and **oxidation** are applied
to these changes (although **oxygen** is **not necessarily
involved**).

- A substance **losing electrons** is said to be **oxidized**.
- A substance **gaining electrons** is said to be **reduced**.

IONIZATION

In most situations, matter takes the form of neutral atoms or molecules. However, when electrons are stripped away or added to neutral atoms, they form charged ions with an overall net positive or negative charge.

IONIZING MATTER

Ions can be **positively** or **negatively charged** and are written down with a **superscript number** after them that indicates their **net charge** – for example, Fe^{2+} or CO_3^{2-}.

Ions are formed by the **addition** or **removal** of **electrons** from an **atom's outer shell**:

- **Negative ions** have **extra electrons** beyond those required for an atom's **neutral state**.
- **Positive ions** have a **deficit of electrons** compared to the **neutral state**.

Positively charged ions are known as **cations**, while **negatively charged** ones are called **anions**.

Ionized atoms are **highly reactive** – in order to regain the **stability** of their **neutral configuration** they will react with **any other substance they come into contact with**.

CREATING IONS

Ions may form in a variety of different conditions:

- In **hot environments** where **collisions** between **fast-moving atoms** or **molecules strip electrons** from their **outer layers** (and **fast-moving free electrons** then **collide** with other **particles**).
- Where **bombardment** with **high-energy electromagnetic rays** (typically **ultraviolet, X-rays** or **gamma rays**) provides enough **energy** for **electrons** to **escape**.
- In **intense electric fields** where powerful **potential differences strip electrons from atoms** to **permit conduction of electricity**.
- As a (generally **short-lived**) **intermediate stage** in **chemical reactions**.

PLASMAS

Often described as a **fourth state of matter**, plasmas are **fluids created by** ions in **isolated environments** that **deny them an opportunity** to **react** and **return** to a **neutral state**. Their **electric charge** and **ability to move freely** mean they can be **shaped** and **influenced** by **magnetic fields** (including those generated by the **flow** of their own **internal electric currents**). **Nuclear fusion reactors** and **particle accelerators** frequently use **powerful magnetic fields** to **contain plasmas for study**.

MASS SPECTROMETRY

Ionization of atoms and molecules is the key to the ingenious technique called mass spectrometry, in which the individual chemical constituents of a sample of matter are split apart like the wavelengths of light in a spectrum.

HOW IT WORKS

A **mass spectrometer** relies on the idea that a **uniform electromagnetic field** will **deflect fast-moving electrically charged particles** by **different amounts** depending on their **mass and charge**:

- Particles with **greater electric charge** are **deflected more**.

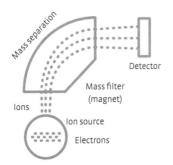

Mass separation

Detector

Mass filter (magnet)

Ions

Ion source

Electrons

- Particles with **greater mass** and **momentum** are **deflected less**.

A typical mass spectrometer **breaks down** a sample of a substance into **electrically charged ions** by **heating it intensely** to form a **gas** and then subjecting it to one of **several processes** that **ionize its atoms** and **molecules** (such as **bombardment** with **electrons** or **X-rays**).

The charged ions are **focused into a beam** before entering an "**analyser**". This may be an **electric** or **magnetic field** that **alters their trajectories** or provides a fixed **accelerating force** (so ions of different masses are **boosted to different velocities**).

Finally, the ions strike a **detector** of some kind that **measures** the **relative proportions of particles** with **different mass-to-charge ratios**, either by the **angle at which they are deflected**, or the **time they take to arrive**.

The result is a "**mass spectrum**" similar to a **spectrum of radiation emission**, in which the **intensity of different peaks** indicates the **proportion of different ions** produced from the **original treated sample**.

APPLICATIONS

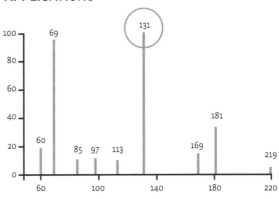

Mass spectrometry offers a useful **shortcut** for **analysing** the **chemical composition of substances**. Importantly, it also **reveals differences in their isotopic composition** that are **not revealed in traditional chemical analysis**.

Applications include:

Radioisotope dating in **archaeology** and **geology** (where the **ratio of isotopes** of a **particular element** indicates the **age** of a sample).

Medical studies of **absorption** into the **human body** (where a patient is given **medicine** or **food** carrying a **specific isotope** that can be **tracked** to **see where it ends up**).

Space exploration: mass spectrometers carried aboard **space probes** allow them to **analyse** the **constituents of interplanetary dust** and the **thin outer atmospheres** of **distant planets**.

INTERMOLECULAR BONDS

Bonds between molecules are generally much weaker than the interatomic bonds within them. They play an important role in holding substances together in their solid and liquid states, affecting their physical rather than chemical properties.

ELECTROSTATIC FORCES

Most **intermolecular bonds** rely on the basic principle of **electrostatic forces** acting to **attract oppositely charged areas** to each other.

While a molecule's **overall electric charge** may be **neutral**, the **distribution of electrons** in the **outer shells** of its **atoms** is often **uneven** (especially in **covalent bonds** where **paired electrons** must **occupy certain regions** in order to be **shared between atoms**). Such molecules are said to be **polar**.

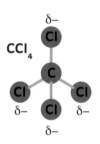

CCl_4

A **concentration of electrons** (and **negative charge**) (marked δ-) in **one part of a molecule** creates a **deficit** (a **net positive charge**, δ+) in **another part** – a so-called **dipole**.

The **opposite ends** of **dipoles** in different molecules **attract** each other, creating **weak attractions**, termed **Van der Waals' forces**, that **hold substances together**.

HYDROGEN BONDING

An atom with an **excess of electrons** due to **chemical bonding** exerts a **particularly strong attraction** to **positively charged hydrogen nuclei** in **surrounding molecules**. This **hydrogen bonding** effect is particularly powerful between H_2O molecules, and **raises** the **melting** and **boiling points** of water significantly.

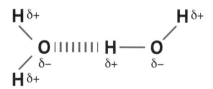

SOLUTIONS

Solutions form when one substance (the **solute**) is **mixed** with another (usually a **liquid**) called a **solvent. Intermolecular forces** from the **solvent** overcome those **holding the solute together**, causing it to **break down** or **dissolve** into **individual molecules**.

IONIC SOLUTIONS AND ELECTROLYSIS

Water's strongly **polar nature** makes it a **particularly effective solvent**. Along with some others, it can overcome the **interatomic bonds** of **Van der Waals' forces**, **breaking a solute apart** into **dissolved ions** (**atoms** or **groups of atoms**) with **opposite electric charges**.

Electrolysis is a method of **separating ionic solutions**. When **electricity** passes through the solution between two **conducting electrodes**, **ions migrate** in **opposite directions** and undergo **chemical reactions** at each **electrode** to form **new substances**.

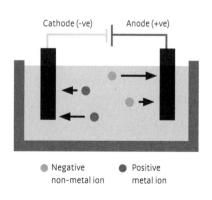

● Negative   ● Positive
non-metal ion   metal ion

X-RAY CRYSTALLOGRAPHY

X-rays are well known for their ability to penetrate soft tissues and image the internal structure of everyday objects – but they also allow scientists to probe the structure of matter on a sub-microscopic scale.

DIFFRACTED RAYS

X-rays have much **shorter wavelengths** than **light rays**, but as **electromagnetic waves** they are **subject to the same phenomena**, including **diffraction** (see p.72).

The **short wavelength** means they are only **diffracted** when they **pass through very narrow gaps**, so the phenomenon is **not seen in everyday situations**.

However, the **surfaces** and **internal structures** of many **molecules** and **crystalline solids** can effectively act as **diffraction gratings** for **X-rays**, causing them to **spread out** and creating **tell-tale diffraction patterns**.

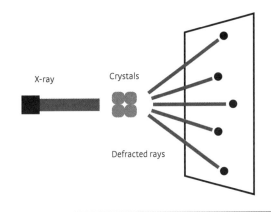

X-ray   Crystals

Defracted rays

FOUNDERS OF CRYSTALLOGRAPHY

In 1912, **Max von Laue** demonstrated **X-ray diffraction** in order to prove that X-rays were indeed a form of **electromagnetic radiation**. He pioneered the idea that diffraction could be used to investigate the **internal structure of crystals**, while father and son **William** and **Lawrence Bragg** worked out **laws** that allowed it to be put into **practical use**.

BIOLOGICAL BREAKTHROUGHS

In the mid-twentieth century, **X-ray crystallography** became a vital tool for **biochemical research**, helping to **unravel the structures** of some of the **most complex organic molecules**.

- In 1951, **Linus Pauling** used it to show that many **protein molecules** have **helical (twisted) structures**.
- Shortly afterwards, **Rosalind Franklin** and **Maurice Wilkins** found evidence that the **DNA molecule** that carries **genetic information** in all **living cells** also has a **tightly defined helical structure**.
- In 1953, **James Watson** and **Francis Crick** used Franklin's research to formulate their **"double helix" model**, explaining that **DNA** has a structure similar to a **twisted ladder**.
- **Dorothy Crowfoot Hodgkin** won the 1964 **Nobel Prize** for her own pioneering work using **crystallography** to **map the structures of molecules** such as **insulin** and **vitamin B$_{12}$**.

CRYSTALS ON MARS

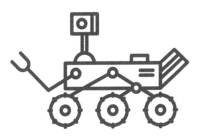

NASA's *Curiosity* **Mars rover**, which arrived on the Red Planet in 2012, carries an **X-ray diffraction experiment** to analyse the **chemical structure of surface minerals**.

ATOMIC FORCE MICROSCOPY

Employing the most delicate form of sensor imaginable, atomic force microscopy uses a physical probe to map out the atomic surfaces of materials. It can even rearrange atoms themselves.

SENSING ATOMS

The principle behind the **atomic force microscope** (**AFM**) is surprisingly simple – it relies on passing a **super-fine, super-sharp probe** back and forth across the surface of a material in order to **scan** it.

The probe bumps its way across the material in response to the **direct mechanical force** from the surface itself or (by **altering the configuration** and **coating the tip with other materials**) forces of **attraction** and **repulsion**, including **Van der Waals' forces** and **magnetism**. This reveals the **structure** and **properties** of the material's surface layer in **atomic-scale detail**.

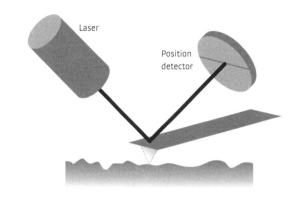

Laser

Position detector

- **Silicon** or **silicon nitride probe** has **tip sharpened** to a **few nanometres** (billionths of a metre) across.
- **Probe** is **mounted** on **one end** of a **cantilever** so it can **move up** and **down**.
- **Laser** directs a **beam of light** onto **reverse of cantilever**.
- **Detector** measures **direction** of reflected **laser beam** and calculates the cantilever's **deflection**.
- **Computer** reconstructs an **image** based on **deflection** at **each point** in the **scan**.

NANOLITHOGRAPHY

AFMs allow not simply the **sensing of individual atoms**, but also their **manipulation**. By generating **electrostatic forces** in the **tip of the probe**, they can **attract** and **repel individual atoms**, and even **pick them up**, **move them around**, and **drop them** – a technique akin to **printing with atoms**, called **nanolithography**.

As a method for **building atomic-scale structures** on a **flat surface**, nanolithography has a variety of applications – most obviously in the manufacture of **integrated circuits** and **related devices**.

In the future, it may be possible to build **three-dimensional nanomachines – complex molecular structures capable of performing specific tasks** – and perhaps even **replicating themselves**.

PUBLICITY STUNT

The **first practical demonstration of nanolithography** took place in 1989, when scientists at **IBM** manipulated thirty-five **xenon atoms** on a **copper surface** in order to **spell out their company's name**.

LIGHT FROM ATOMS

Materials can emit light in a number of different ways – most importantly, through the vibration of their individual atoms and molecules, and through changes to the internal structure of these particles.

MATERIALS IN MOTION

Thermal radiation is a **release of electromagnetic radiation** (as **visible light** or other forms) caused by the **heating of bulk materials**.

Within these materials, **complex electrostatic bonds link each particle to its neighbours**. These **fields** may be due to **strong chemical bonding** or **weaker forces** caused by the **uneven way that electric charge is distributed within molecules**.

When the **kinetic energy** of **atoms** and **molecules** is **boosted by heating**, the **bonds** between them are **stretched** and **distorted**, creating an **oscillating electromagnetic field** throughout the bulk of a material.

The oscillations release **radiation** at a **range of wavelengths** that are related to the **material's temperature** (itself an indication of the molecules' **kinetic energy**).

ATOMIC EMISSIONS

Emission-spectrum radiation is generated when individual **atoms** or **molecules absorb** and **release energy**.

The process usually involves an **electron particle** inside the atom being **excited by some external process**, allowing it to **switch briefly** to a **higher energy level** before **falling back to its original state** (see p.112).

In contrast to **thermal radiation**, emission-spectrum radiation is **limited** to a **range of specific narrow wavelengths** dependent on the **materials being excited**. The science of **spectroscopy** involves **studying these wavelengths** to **learn more about their source material**.

COLOUR AND TEMPERATURE
Human vision interprets the various **wavelengths** emitted by **luminous objects** as a **single overall colour**, but it's easy to see how the **radiation changes** to **shorter wavelengths** and **higher energies** as a **metal bar is heated**:

Temperature	Visible colour
580°C	Dull red
930°C	Bright orange
1400°C	White hot (combining lower-energy oranges and yellows with higher-energy greens and blues)

BLACK-BODY RADIATION

Thermal radiation from hot materials mimics a distinctive pattern know as the black-body radiation curve, linking the amount of radiation and its spread of wavelengths and colours to the material's temperature.

WHAT IS A BLACK BODY?

A **black body** is an **ideal object** that is a **perfect absorber** of all **radiation** regardless of **frequency** or **angle of approach**. A black body can emit **radiation of its own**, but its surface is completely **unreflective**. The concept is useful because it **simplifies the models** necessary to describe **thermal radiation**.

In reality, **incandescent objects** such as **stars** tend to be **grey bodies** – they emit a **fraction** of the **thermal energy** they would if they were **truly black**, but still **follow most rules of black-body behaviour**.

PERFECT BLACK BODIES

A **real-world approximation to black bodies** is the **cavity radiation source** – a **hollow sphere** with a **blackened interior** and a **small entrance hole**. Any **radiation entering the cavity** is very **unlikely to find its way back out**, making it a **near-perfect absorber**. Modern materials including **"super black" paint** and **coatings of carbon nanotubes** can do **even better**.

TEMPERATURE, COLOUR, AND ENERGY

When black bodies produce **thermal radiation**, the **wavelength** and **intensity** emitted follows a **distinctive curve rising to a central peak**. The **higher** the body's **temperature**, the **stronger** the **peak** and the **shorter** its **wavelength**.

By **comparing its radiation** at a few **specific wavelengths**, it's possible to work out the **temperature** of a black body (or at least the **effective temperature** of a **real-world grey body**).

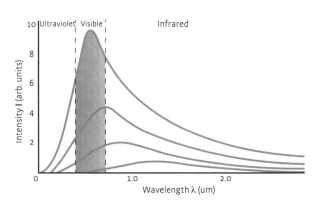

THE STEFAN–BOLTZMANN LAW

The **total energy radiated by a black body** is **strongly governed** by its **temperature**, and described by the Stefan–Boltzmann law:

The energy radiated per square metre of a black body's surface each second, $j*$, is given by:

$$j* = \sigma T^4$$

where T is the body's **absolute temperature** (measured in kelvin above absolute zero – see p.86) and σ is the **Stefan–Boltzmann constant**, 5.67×10^{-8} W m^{-2} K^{-4}.

SPECTROSCOPY

Spectroscopy is the study of light emitted or absorbed by different forms of matter in order to identify their constituent elements and learn more about their physical properties.

THE SPECTROSCOPE

Most **natural processes** that **emit** or **absorb radiation** do so **preferentially** – they either involve a **broad but still limited range of wavelengths** (as in **thermal radiation**) or a number of **very specific wavelengths** (as with **emission spectra**).

A spectroscope **splits light up** into a **broad spectrum** using a **diffraction grating** (see p.72) that **disperses different wavelengths** at **different angles**. The **changing intensity of light** depending on **colour** and **wavelength** can be observed through an **eyepiece** or recorded on an **electronic sensor**.

TYPES OF SPECTRA

Spectra fall into three broad categories:

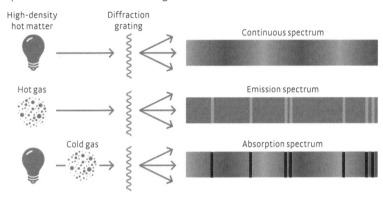

- **Continuum**: A **rainbow-like distribution** across a **broad range of wavelengths**, typical of **black-body radiation**. The **intensity** of **different colours** can indicate the **temperature** of the **source**.
- **Emission**: Generally **dark** except for a series of **brightly coloured lines** at **specific wavelengths**, caused by the **relaxation** of **excited atoms** in the **source material**.
- **Absorption**: A **continuum spectrum** overlaid by **narrow dark lines**. Absorption spectra are the **flipside of emission** – they form when **atoms absorb energy** from a **continuum source** at **specific wavelengths** in order to become **excited**.

SCIENCE FROM SPECTRA

Spectra can reveal a huge amount about materials in the **laboratory** and the **wider Universe**. For instance:

- Wavelengths of **emission** and **absorption** linked to specific **elements** reveal the secrets of their **inner structure**, but also act as **unique spectral fingerprints** to identify the **same elements elsewhere** (for instance, in **distant stars**).
- **Unexpected lines** can **predict new discoveries** – and even **new elements** (**helium** was found this way in 1868).
- **Shifts** in the **expected wavelength of spectral lines** can reveal when objects are **in motion** (and affected by the **Doppler effect**) or subject to **intense gravity** or **strong magnetic fields**.

TYPES OF WAVE

Waves are disturbances that transfer energy from one place to another – usually (though not always) through a transmitting medium such as air or water. Their behaviour is governed by very different rules from the classical mechanics of particles.

TRANSVERSE WAVES

The most **familiar** and **easily understood** waves are those in which the **physical disturbance of the medium** happens at **right angles** to the **direction of the wave's overall motion**. In transverse waves such as **water waves**, the disturbed particles themselves **do not change position significantly** in the **direction of the wave's motion**, but their **temporary disruption** in the **transverse** direction **passes on the energy** to their neighbours.

Examples include:
- water waves
- "secondary" seismic waves
- light waves

LONGITUDINAL WAVES

Many **natural waves** are **longitudinal** in nature – a **back-and-forth disturbance** that happens in a direction that is **parallel to the wave's overall motion**. Longitudinal waves are **successive areas** in which the medium is **compressed** and **rarefied** (**spread out**).

Examples include:
- sound waves
- "primary" seismic waves

SHARED CHARACTERISTICS

All **waves** are defined by a **basic set of characteristics**:

- **Wavelength**: The **interval** between successive "**peaks**" or "**troughs**".
- **Frequency**: The **rate** at which waves **move past** a particular **arbitrary point** (measured in **Hertz**).
- **Amplitude**: The **strength** of the **disturbance** associated with the wave, **relative to** the **equilibrium state** of the medium.
- **Wavenumber**: Useful for some **calculations**, this is simply the **number of waves** fitting into a particular **unit of distance** (1/wavelength).

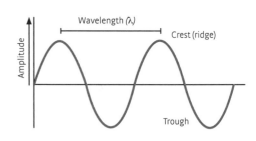

SINE WAVES

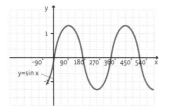

Many waves in **nature** follow the form of a **simple mathematical sine wave**, in which the **"restoring" force** increases with **displacement** from the **equilibrium**. The wave displacement **slows** as it approaches its **maximum**, while **motion** in the wave is **fastest** as it **passes** (and **overshoots**) the **equilibrium**.

INTERFERENCE

When waves overlap each other, the individual disturbances they cause are added to each other, reinforcing in some places and cancelling out in others. This phenomenon played a key role in discovering the wave-like nature of light.

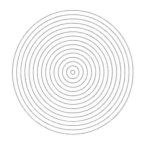

WATER RIPPLES

The effects of **interference** are easy to see in **water** – throwing two stones simultaneously into the same pond sets up **two patterns** of **spreading concentric ripples** with **alternating peaks and troughs**.

Where two peaks or troughs **cross over each other**, the overall **height** or **depth** of the wave is **increased**, but if a peak and a trough **coincide**, they **cancel out** to leave a **flat surface**.

RIPPLE TANKS

Around 1800, physician **Thomas Young** invented a **shallow water tank** that can be used to **study** many **different wave behaviours**. Ripple tanks are still widely used as a **teaching aid in schools**.

YOUNG'S BREAKTHROUGH
- **Huygens' model**: Waves are present **only** where they **perfectly reinforce**.
- **Young's model**: Waves are present **everywhere** except where they **perfectly cancel out**.

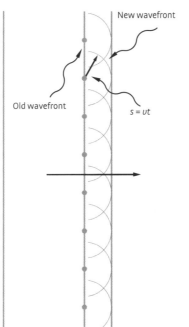

New wavefront

Old wavefront

$s = vt$

HUYGENS' PRINCIPLE

Dutch physicist **Christiaan Huygens**, who argued that **light was a wave**, proposed a useful way of **modelling various optical** (and other wave) **effects** in 1690:

- **Each point** on an **advancing wavefront** is a source of new "**wavelets**" spreading in **all directions**.
- The overlapping wavelets **interfere** and **cancel out** in **nearly all directions**.
- The **line** along which overlapping wavelets **reinforce each other** dictates the **direction** in which the **wavefront moves**.

Interference

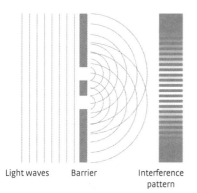

Light waves Barrier Interference
pattern

THE DOUBLE-SLIT EXPERIMENT

In 1800, **Young** proved the **wave-like nature of light** by showing that **two beams** of light, **spreading out** after they passed through **narrow slits** (an effect called **diffraction** – see p. 72), **interfered** with each other where they **overlapped** to create **complex patterns** on a screen.

SOUND WAVES

Sound is a longitudinal wave that travels through air and other materials as regions of compression and rarefaction. We experience it, with various senses, as a form of pressure wave – most notably through hearing.

SOUND PROPERTIES

Sound has **typical wave properties** of **speed**, **frequency**, **wavelength**, and **amplitude**. Despite being a **longitudinal wave**, it is often **easier to visualize these properties** on a **transverse wave diagram**.

A sound's **intensity** is the **power it can deliver** per unit area, measured in **watts per square metre**. It is defined by the **equation**

$$I = pv$$

where p is the **pressure difference** created by the wave, and v the **velocity of moving particles** in the medium.

Because sound waves usually **spread spherically across space**, **intensity falls off** in **inverse proportion** to the **square of distance** d:

$$I\,(d) \propto 1/d^2$$

SOUND LEVELS

The **strength** of sound is usually described in terms of **sound intensity level**, a different property related to **how we perceive sound**, defined as:

$$\beta = 10 \log_{10} (I/I_0)$$

with I_0 as a **reference intensity** of 10^{-12} W/m² (the **lower limit** of **human perception** at a **frequency of 1,000 hertz**).

β is measured in **decibels**. The **logarithmic scale** means that a **difference of 10 decibels** indicates a *factor of 10* **difference in intensity**.

STANDING WAVES AND HARMONICS

Our intuitive image of a wave imagines it going somewhere – transferring energy through space. But if the wave medium is confined in some way, it can turn back on itself, resulting in the formation of standing waves.

HARMONICS

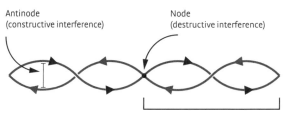

Antinode
(constructive interference)

Node
(destructive interference)

▶ = Wave emitted from source
▶ = Reflected wave

One wavelength

When the medium through which a wave can travel is **confined**, it can only support waves of **certain wavelengths**, known as **harmonics**.

- When a **harmonic wave** reaches the **end of the medium** and **reflects**, its reflection continues the **original wave pattern**, resulting in a wave that is a **precise mirror image of the original**.
- Within the **combined standing wave**, some points known as **nodes** will **not move at all**, while others will **oscillate up and down** or **back and forth**.
- **Non-harmonic waves** with **different lengths** will rapidly **break down** due to **interference** as they **reflect back on themselves**.

MUSICAL HARMONIES

According to a famous Greek legend, the philosopher **Pythagoras** discovered the **basic pattern of harmonics** after hearing the **different tones** caused by **hammers of different weights** striking an **anvil**. In reality the experiments of Pythagoras and his followers used **plucked strings** of **different lengths and weights**. The **interval** between **two harmonics** is known as an **octave**, and can be **bridged** by a sequence of **intervening frequencies** to make a **musical scale**.

HARMONIC WAVELENGTHS

Standing harmonic waves have **wavelengths** that are **simple ratios** of the **length of the oscillating medium** L.

MUSICAL WAVES

The key to the **infinite variety of music** lies not just in the **frequency** and **amplitude** of the **waves produced**, but also in a quality called **timbre** – the **complex shape** of the wave produced by different **musical instruments**, **voices**, and **styles of playing**.

Perfect **sinusoidal sound waves** can only be produced by **electronic synthesizers**.

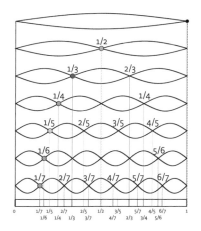

	W*	F*
First harmonic (fundamental tone):	$2L$	f
Second harmonic (first overtone):	L	$2f$
Third harmonic (second overtone):	$2L/3$	$3f$
Fourth harmonic (third overtone):	$L/2$	$4f$
Fifth harmonic (fourth overtone):	$2L/5$	$5f$

*Wavelength *Frequency

THE DOPPLER EFFECT

Today we're all familiar with the Doppler effect from the shift in the pitch of sound when an emergency siren zooms past. For astronomers and other physicists, however, Doppler shifts are an important tool.

HOW IT WORKS

The Doppler effect is a **simple consequence** of the **relative motion** of **source** and **observer**:

- **Sound waves** spread out in **all directions** at the **same rate**.
- If the **observer** and **source** move **towards each other**, the observer **passes successive peaks** at an **increased rate** and so measures a **higher frequency** and **shorter wavelength**.
- If the **observer** and **source** move **away from each other**, the observer's **crossing of each successive peak** is **delayed**, so they measure a **lower frequency** and **longer wavelength**.

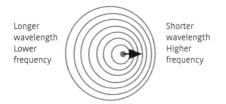

Longer wavelength
Lower frequency

Shorter wavelength
Higher frequency

DOPPLER'S PREDICTION

In 1842, Austrian physicist **Christian Doppler** pointed out that the **frequency of waves** passing a **fixed observer** must **change slightly** depending on whether the **source** of the waves is **moving towards** or **away from them**. He **wrongly thought** this might explain the **different colours of stars**.

In 1845, Dutch scientist **C. H. D. Buys Ballot** experienced the Doppler effect for **sound** by putting **musicians on a railway carriage** and getting them to play a **steady note** as they **sped past**.

RED AND BLUE SHIFTS

The **speed** and **frequency** of **light waves** are so **high** that the Doppler effect in **light** is far **too small** to affect the **overall perceived colour of objects** in any but the most **extreme** situations. However, **astronomers** have used the Doppler effect since the nineteenth century by measuring the **displacement of spectral lines** (the **chemical fingerprints** of **specific elements** in starlight, which occur at very specific **wavelengths** and **frequencies**). Light from **retreating objects** is **"red shifted"** to **longer** wavelengths, while light from **approaching ones** is **"blue shifted"** to **shorter** wavelengths.

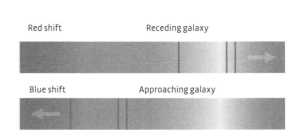

Red shift   Receding galaxy

Blue shift   Approaching galaxy

ELECTROMAGNETIC WAVES

The discovery of diffraction and interference patterns in beams of light showed behaviour that could only be explained if it was a wave. But the true nature of that wave did not become clear until the 1850s.

MAXWELL'S THEORY

In 1862, **James Clerk Maxwell** provided evidence that light was an **electromagnetic wave** (a moving disturbance caused by **transverse electric** and **magnetic waves** at **right angles** to both **each other** and their **direction of movement**):

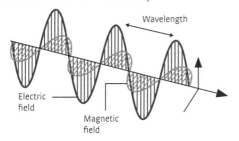

Wavelength

Electric field

Magnetic field

- Maxwell's discovery was triggered by **new discoveries** in the area of **electromagnetism** – especially the realization that **magnetic fields** can **alter** the **polarization of light**.
- The electric and magnetic components **alternate** so that the magnetic field is **zero** when the **electric disturbance** is at its **greatest** and **vice versa**. This allows the wave to **reinforce itself**.
- Working from the **natural constants** that describe the **strength** of **magnetic and electric fields**, Maxwell calculated that the **speed** of **electromagnetic waves** in a **vacuum** would be about the **same as the speed of light** – now known to be 299,792 km/s (the speed of light, c).

THE ELECTROMAGNETIC SPECTRUM

Because Maxwell's waves move at a **constant speed**, their **frequency** and **wavelength** are **inextricably linked** together by the **equations**

$$\lambda = c/f \qquad f = c/\lambda$$

higher frequency f = shorter wavelength λ
lower frequency f = longer wavelength λ

Maxwell predicted that the **range** of his **electromagnetic waves** could extend **far beyond the limits** of the **visible, infrared, and ultraviolet radiations** known at the time. **Heinrich Hertz's** discovery of **radio waves** in 1886 **confirmed** that his **theory was right**.

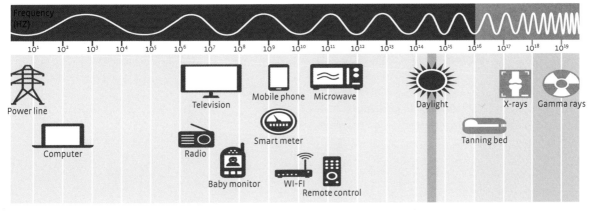

Frequency (HZ)

10^1 10^2 10^3 10^4 10^5 10^6 10^7 10^8 10^9 10^{10} 10^{11} 10^{12} 10^{13} 10^{14} 10^{15} 10^{16} 10^{17} 10^{18} 10^{19}

Power line | Computer | Television | Radio | Baby monitor | Smart meter | WI-FI | Mobile phone | Microwave | Remote control | Daylight | Tanning bed | X-rays | Gamma rays

REFLECTION

Reflection is the tendency of a wave to bounce off a surface and continue moving in a direction related to its angle of approach. It is especially familiar from light, but it applies to all types of waves.

EUCLID'S MODEL

In order to understand **reflection**, Greek mathematician **Euclid** defined a concept called the **angle of incidence**.

The angle of incidence is the **angle between** the **incoming light wave** and the "**normal**" – an **imaginary perpendicular** to the **mirror's surface** at the **point** where the **light ray strikes**.

A **mirror** reflects light at an **angle of reflection equal to its angle of incidence**, and on the **opposite side** of the **normal**.

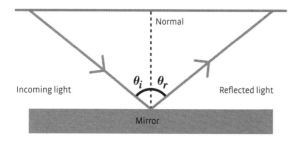

HERO'S PRINCIPLE

Greek mathematician **Hero of Alexandria** realized that **Euclid's laws of reflection** meant that reflected light would **always follow the path of least distance**. Much later this was generalized by **Pierre de Fermat** to explain other **optical phenomena** in terms of the **path** along which the light takes **least *time* to travel**.

Both are examples of a "**least action principle**" – the **tendency of processes in physics** to follow the **most efficient pathway possible**.

ABSORPTION, SPECULAR AND DIFFUSE REFLECTION

The **quality of the surface** that a **wave strikes** plays a key role in **how reflection takes place**:

- If the reflecting surface is composed of material that **absorbs the energy of the incoming wave** in some way, the **amount of energy reflected**, and **strength of the reflected wave**, are **reduced**.
- If the reflecting surface has a **smooth surface** that **does not absorb the energy of incoming waves**, then the reflection will **preserve** most of the "**structure**" of the **incoming wavefront** – parts of the wave that were **alongside each other** in the **incident wave** will **remain alongside each other** when reflected.
- If the reflecting surface is **rough** on a **scale comparable** to the **size of the incoming waves**, then the wave pattern will be **disrupted** during reflection, resulting in a **diffuse reflection** in which the **structure of the incoming wave is lost**.

SPECULAR REFLECTION

DIFFUSE REFLECTION

REFRACTION

Refraction is the tendency of waves to change their speed and direction as they move from one medium to another. This is due to the different amounts of energy required for the wave to propagate through the different materials.

SNELL'S LAW

When a wave enters a medium that **transmits it more slowly than before**, it naturally tends to **bend towards the normal** (the line **perpendicular to the surface** at the **point of entry**).

When it enters a medium through which the wave can move **more quickly** than before, it **bends away from the normal**.

The **angle of incidence**, **angle of refraction**, and **wave speed** are **governed by a rule** called **Snell's law**:

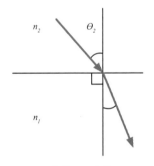

$$n_1 \sin \Theta_1 = n_2 \sin \Theta_2$$

In the case of **light**, the **speed of waves** in a particular **transparent material** is known as its **refractive index**. A **vacuum** has a **refractive index** of 1 – because the speed of light in **other materials** is **always slower** than this, their refractive index is always **less than 1**.

HUYGENS' MODEL OF REFRACTION

Christiaan Huygens' model of "wavelets" makes it easy to understand what happens during **refraction**:

- Unless the **wavefront** approaches the boundary surface at **right angles**, one side will **encounter the new medium** while the others are **moving through the original one**.
- Along the **leading edge** of the wavefront, **wavelets** propagating from **points that have entered the new medium** will **spread at a different speed** from those that are **still in the original medium**.
- The **line** along which wavelets **reinforce** will therefore **bend** either **towards** or **away from** the **normal**.

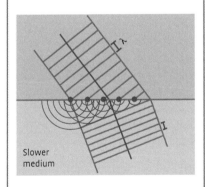

Slower medium

PRISMS

Because it is linked to the **energy** required to **carry a wave** through **different media**, refraction affects **waves** with **different energies** (measured by their **frequencies**) by **different amounts**.

For **light**, this means that **shorter-wavelength blue colours** are refracted **more** than **longer wavelengths** towards the **red end of the spectrum** – an effect called **dispersion**.

A **wedge-shaped prism** exaggerates the dispersion effect to create a broad **rainbow-like spectrum**.

DIFFRACTION

Diffraction is the tendency of waves to spread out after they pass through a narrow gap, and to "spread" around the edge of barriers.

HOW IT WORKS

Diffraction is **one of the most characteristic** wave properties. It arises because the **energy causing displacement** on the very **edge of a barrier** is **hard to contain** – it has a tendency to **interact with the barrier** in ways that make it **spread out**.

- In **Huygens' wavelet model**, **diffracted waves** are simply those formed by **wavelets** generated at the very **edge of the wave** as it passes through a **slot in a barrier**. These **naturally spread** into the **undisturbed space** in the barrier's "**shadow**".
- In fact, passage through a slot gives rise to **complex interference effects** that **don't simply spread its effects into the shadow area**, but also **change the pattern** of the **wave itself**, **reinforcing** in some places and **cancelling out** in others.
- The **strength** and **details** of the **diffraction effect** are influenced by both the **width of the slot** and the

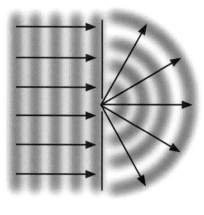

Incident plane
wave

Wave past
small opening

wavelength of the waves passing through it – as a general rule, **longer wavelengths** are diffracted **more** than **shorter ones**, and the effect is at its **strongest** when the **aperture** of the slot is **close to the wavelength** of the waves involved.

DIFFRACTION GRATINGS

A **diffraction grating** is a **series of fine slits** or a **scored reflective surface** that **disperses different wavelengths of light** by the principles of diffraction. Gratings were perfected in the nineteenth century by **Joseph von Fraunhofer** and others – the **spread-out spectra of light** they created revealed **fine details** that could not be seen through **prisms**, and gave rise to a **new science** called **spectroscopy** (see p.63).

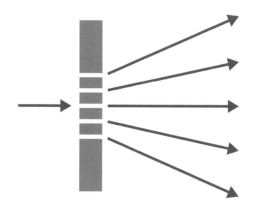

WAVES

SCATTERING

Waves can lose energy or change direction in several ways as they interact with their environment. These effects, known as scattering, are particularly important in the study of electromagnetic waves such as light.

RAYLEIGH SCATTERING

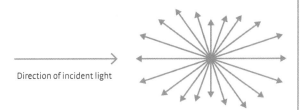

Direction of incident light

This common form of scattering occurs when **light interacts** with **particles** that are **significantly smaller** than its **own wavelength**:

- Particles such as **air molecules** have an **internal arrangement of electrons** that can become **polarized**.
- The **electromagnetic field** of a **passing light ray** causes the particle's **polarization** to **oscillate** at the **same frequency**.
- The **molecule** therefore **emits radiation** of its own, but the **original light wave** continues **unaffected** except for a **change of direction**. This type of scattering is therefore said to be **elastic**.

In most situations the **strength of scattering** is **inversely proportional** to the **fourth power of the wavelength** (i.e. it increases as λ^4 gets smaller). In **Earth's atmosphere**, this means that **blue light** is **scattered far more strongly** than **red**, and this is why the sky itself appears blue.

RAMAN SCATTERING

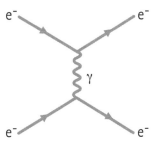

Although similar in many ways to **Rayleigh scattering**, the **Raman effect** is **inelastic** – it usually involves the wave **losing energy** and **shifting** to a **longer wavelength**. Raman scattering occurs when a light wave **strikes a particle head-on**: some of its **energy** is **transferred** to **vibrations in the molecule**, before it is mostly **re-emitted**.

COMPTON SCATTERING

A third type of scattering occurs when **electromagnetic waves interact with charged particles** such as **electrons**. This scattering **injects energy** into the electron and produces a **scattered wave** with a **longer wavelength**.

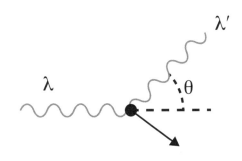

POLARIZATION

Polarization is a special property of light waves – it refers to the plane in which the transverse wave is oscillating, which may be aligned in any direction perpendicular to the direction of travel.

TYPES OF POLARIZATION

- **Unpolarized light**: Light waves vibrate in **many different planes**.
- **Plane polarized**: Light waves are vibrating in a **single fixed plane**.
- **Circular polarized**: The **polarization plane** is **twisted**, **rotating** as the waves **move through space**.

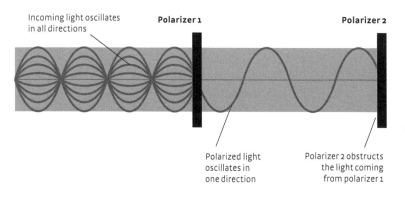

Incoming light oscillates in all directions

Polarizer 1

Polarizer 2

Polarized light oscillates in one direction

Polarizer 2 obstructs the light coming from polarizer 1

BIREFRINGENT CRYSTALS

Birefringent crystals found in nature have structures that give them different **refractive indices** depending on the **polarization** of the **light passing through them**. This means that **components** of **unpolarized light** vibrating in **different planes** pass through them on **different paths** to create a **double image**.

- **Viking sailors** are thought to have used birefringent "**Iceland spar**" to observe **natural polarization patterns** in the **sky** as a **navigation aid**.

SUNGLASSES AND LIQUID CRYSTALS

Modern **Polaroid filters** use long **polymer molecules** to create an **absorptive polarizing filter**. The molecules can be **embedded** in **thin films** to create a **permanent polarizing screen** (as in **sunglasses**), or **suspended** in **liquid form** so that their **alignment changes** when an **electric current** is applied (the principle behind **liquid crystal displays**).

MAKING POLARIZED LIGHT

Polarized light **can be created in several different ways**, both **naturally** and **artificially**:

- **Birefringence**: Passing an **unpolarized beam of light** through a **birefringent crystal** creates **two beams** polarized at **right angles** to each other.

- **Reflection**: If unpolarized light **strikes the boundary** between **two transparent media** with **refractive indices** n_1 and n_2 at **Brewster's angle**:

$$\theta_B = \arctan\left(n_2/n_1\right)$$

light of one polarity is **transmitted perfectly** through the **surface** while light of the **other polarity** is **entirely rejected**, creating a **polarized reflection**.

- **Absorption**: A **filter** with a **fine structure** of **parallel opaque lines** running in a **single direction** so that **only light of one polarization can pass through**.

CARRIER WAVES

Many modern technologies rely on signals carried by electromagnetic waves (particularly radio waves). The process of manipulating a radio wave in order to embed a signal within it is known as modulation.

AMPLITUDE AND FREQUENCY MODULATION

The **AM** and **FM modulation systems** change the properties of an **analogue (continuously varying) carrier wave** in different ways in order to **carry information**:

Amplitude modulation: The **intensity** of a **high-frequency carrier wave** is **varied** to **reflect** the **waveform** of a **lower-frequency wave** carrying the **signal information**.

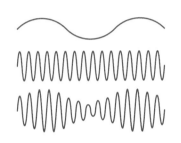

Frequency modulation: The **frequency** of the **high-frequency carrier** is **adjusted** to **reflect variations** in the **underlying signal data**.

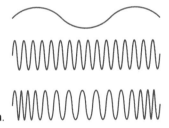

FOURIER TRANSFORMS

In order to **extract** the **original signal data** from a **modulated radio wave**, engineers rely on a **mathematical tool** known as a **Fourier transform**:

$$\hat{f}(\xi) = \int_{-\infty}^{\infty} f(x)\, e^{-2\pi i x \xi}\, dx$$

The **maths** looks daunting, but in essence, a Fourier transform **breaks down any signal** that **varies over time** into a **series of overlapping patterns** with **different frequencies** – rather like identifying the **individual notes** in an **orchestral chord**.

DIGITAL SIGNALS

Modern **electronic processing** allows **analogue wave signals** to be transmitted in **digital form** rather than by **traditional modulation**:

- The **value** of the **analogue signal** is continually **measured** or **sampled**.
- This value is **rounded** to the **closest approximation** on a **numerical scale of signal strength**.
- This **number** is **converted** into a **stream of binary digits** (0s and 1s) and transmitted in a **stream of data pulses** known as **bits of information**.
- The **binary data** in used to **reconstruct** the **original signal** at the **receiver**.

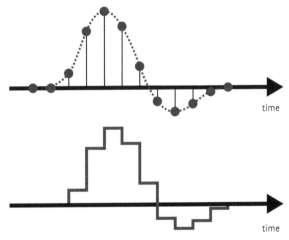

WHY BINARY?

In binary, a **string of eight bits** (a **byte**) can represent **any value from 0 to 1,023**; **two bytes** can cover **values up to 65,535**, and n **bytes up to 2^n-1**. The digital approach **sacrifices some nuance** from **analogue originals** in exchange for **clarity of transmission**, but **sampling rates** and **ranges of values** are now **so high** that the **difference is unnoticeable**.

OPTICAL INSTRUMENTS

Light is one of the most important ways in which we learn about the Universe around us, and a variety of optical instruments have been invented and developed to reveal more through magnified or otherwise improved images.

SIMPLE LENSES

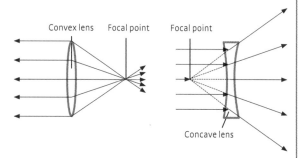

Convex lens   Focal point   Focal point

Concave lens

A **lens** is a piece of **glass** curved in such a way that it **bends light rays** striking it **towards** (or sometimes **away from**) each other:

- The **curvature** of the surfaces and the **thickness** and **refractive index** of the glass **control the paths** taken by **refracted light** as it **enters** and **leaves** the lens.
- **Convex lenses** bend light onto **converging paths** that **cross** at a "**focal point**" before **spreading out again**. Light rays that cross are said to form "**real**" images.
- **Concave lenses** bend light onto **diverging paths**. These rays never **cross over**, but appear to be **spreading from a focal point behind the lens**, forming a "**virtual**" image.
- For ease of comparison, a lens or optical system's **optical power** is usually described in terms of its **focal length** – the **distance** at which it will bring **perfectly parallel light rays** to a **focus**.

RESOLVING POWER

A crucial concept in optics is **resolution** – an instrument's ability to **distinguish between two points separated by a small angle**. In practice this varies depending on an **observer's eyesight**, but a standard test is to model the **potential overlap** of "**Airy disks**" – the **concentric diffraction patterns** created around a **point of light** in even the most perfect optical system:

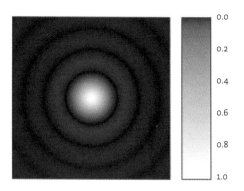

The **angular radius** θ of an Airy disk formed by a **single lens**, from its **centre** to the first "**null**" (**dark ring**), is given by:

$$\theta = 1.22 \; l/D$$

(where D is the lens diameter in metres). Resolution is determined by the **size of the disk**, so, as a rule, **larger lenses** offer **better resolution**, and **shorter wavelengths** produce **sharper images** than **longer ones**.

TELESCOPES

The invention of the telescope transformed astronomy and, through the discoveries of Galileo, began a scientific revolution. Centuries later, telescopes have changed almost beyond recognition.

TELESCOPE BASICS

Telescopes come in two basic types: **lens-based refractors** and **mirror-based reflectors**. In both cases, the **optical elements** are designed to **focus parallel light rays** coming from sources at a **great distance**. A telescope's twin tasks are to **gather more light** than the **human eye alone can see**, and **produce a magnified image**.

Refractors gather light with a large **front lens** called the **objective**, and **bend it** to a **focus point** in a **sealed dark tube**. As the light rays **diverge**, they are **intercepted** by a **smaller lens** called the **eyepiece**, which **bends them back** onto a **near-parallel path** for **observation**.

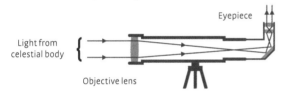

Reflectors collect **light rays** using a **curved mirror** at the **rear** of the telescope rather than a **lens** at the **front**. The light rays follow **converging paths** and are **reflected again** off a **secondary mirror** towards an **eyepiece lens** that may be **mounted** at the **side** of the telescope (in the **Newtonian design**) or at the **back**, accessible through a **hole in the primary** (in the **Cassegrain design** and its relatives).

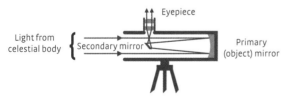

The **magnification** M of any telescope is given by the ratio of **focal lengths** (f) of its elements:

$$M = f_{primary/objective} / f_{eyepiece}$$

PROS AND CONS

Different telescope designs have their own **advantages** and **problems**:

- Telescopes with a **single concave primary lens** are prone to **coloured fringes** called **chromatic aberration**, though this can be avoided using **compound lenses**.
- A **typical primary mirror** has about **half the resolving power** of a lens with the same diameter.
- Keeping down the **weight** of **large lenses** requires **thin curvature** and a **very long focal length** requiring an **impractically long telescope tube**.
- In contrast **large reflecting telescopes** keep their **weight down** with relatively **thin mirrors**, while their **folded light paths** result in more **compact instruments**.

GIANT TELESCOPES

The **largest modern telescopes** are all **reflectors**, often with **giant mirrors** of 10 metres diameter or even more that are made from an **array of hexagonal segments**. **Computer control** allows the **shape** of the mirror to be **precisely adjusted** to account for **distortions** and even **atmospheric turbulence**.

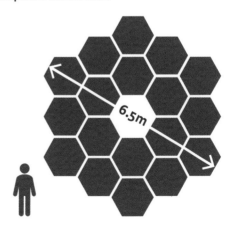

6.5m

MICROSCOPES

While telescopes focus light rays from a great distance, microscopes face a different challenge –
producing a magnified image using the spreading light rays from small objects at close proximity.

WAVES

MAGNIFYING GLASSES AND SIMPLE MICROSCOPES

The traditional **magnifying glass** is a **single lens** with at least one **convex surface**. **Refraction** in the glass **bends diverging light rays** from the object onto paths that are **closer to parallel**, so they form a **larger image** in the eye.

The magnifying glass has a **major limitation**: achieving **stronger magnification** or viewing a **larger area** of the object requires a **thick, sharply curved lens** prone to **distortion**.

In the 1650s, **Antonie van Leeuwenhoek** developed an **advanced simple microscope** by using a **small spherical glass bead** as a **lens**, mounted on a framework that **supported** and **moved** the **object** he **wished to study**. The **field of view** was **tiny**, but **magnifications** could exceed **270 times**.

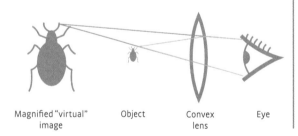

Magnified "virtual" image   Object   Convex lens   Eye

COMPOUND MICROSCOPES

The **compound microscope** design uses **two lenses**, and was invented in the **early 1600s** at around the same time as the **telescope**.

- A small **diameter** of **objective lens** limits the **angular spread** of **light rays** from the sample, allowing it to remain **relatively thin** and **weak**.
- This lens brings light rays to a **focus point** inside the **microscope tube**, from which they **diverge**.
- An **eyepiece** larger than the **objective** then **bends the light** to create the **enlarged image**. Advanced designs add a **third intermediate lens** to bend the light **more strongly**.

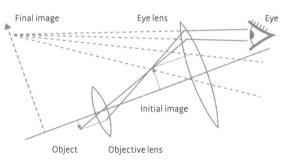

Final image   Eye lens   Eye

Initial image

Object   Objective lens

ILLUMINATION PROBLEMS

Because they **gather light** from just a **small part** of an object, microscopes often **struggle with illumination**. **Robert Hooke** focused **candlelight** onto his subjects with a **water-filled glass bulb**, while **van Leeuwenhoek** backlit his subjects with **light from the sky**. It was only in the nineteenth century that **more intense methods of illumination**, along with **high-contrast dyes** to **highlight details**, allowed the **compound microscope** to **achieve its real potential**.

INTERFEROMETRY

Interferometers are devices that deliberately create interference between waves, allowing information to be extracted and very precise measurements to be made in a wide range of different scientific fields.

INTERFEROMETER BASICS

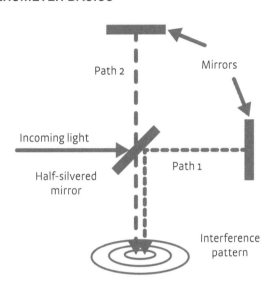

Path 2

Mirrors

Incoming light

Half-silvered mirror

Path 1

Interference pattern

APERTURE SYNTHESIS

Astronomers use a **type of interferometry** known as **aperture synthesis** to observe detail **beyond the reach of any single telescope. Minute variations** in the **path length** of **radiation** arriving at two or more **linked telescopes** can be used to calculate the **exact distribution of radiation coming from the object**. The technique is **simplest** and **most effective** for **radio waves**, but is increasingly used with **visible light**.

- A **single wave** such as a **beam of monochromatic (single-wavelength)** or **laser light**, passes through a **beam splitter** (such as a **partially silvered mirror**) that sends it along **two different paths**.
- The beams are **reunited** so that the **overlap** between **successive peaks** and **troughs** creates an **interference pattern**.
- The **path length** of **one** or **both beams** is **modified** along their journey by the **phenomenon under investigation – changes** to the **interference pattern** reveal the **details of this modification**.

APPLICATIONS

- In **medicine**, **optical coherence tomography** creates **high-resolution images** of **tissue structures** by **passing one beam through a sample** and **combining** it with a **beam reflected off a plane mirror**.
- The **LIGO gravitational wave detector** uses interferometry to **detect tiny distortions** in the **scale of space** itself as **light reflects back and forth** within **two-kilometre-long perpendicular tunnels**.
- The **Michelson–Morley experiment** was a famous **"failed" interferometry experiment** aimed at **measuring changes** in **path length** due to **differences** in the **speed of light**.
- **Newton interferometry** tests the **accuracy of surfaces** being **machined into precise shapes – different shapes** give rise to **different patterns of interference fringes**.

HOLOGRAPHY

Holography is a technique that allows the storage and reconstruction of images carrying far more information than traditional photography. It also has applications far beyond making three-dimensional images.

HOLOGRAPHIC IMAGING

While a **normal photo** records the **intensity of light** from an **object** or **scene**, **projected** as an **image** onto **film** or a **CCD sensor**, a **hologram** preserves the **three-dimensional "light field"** around the object, which can be **observed from different angles**.

Holograms use **light-sensitive film** to preserve an **interference pattern**, created using **interferometry**. **One-half** of a **split laser beam** is used to **scan** the **target object**, while the **other half** is **bounced off a plane mirror** before the two are **recombined**.

Viewing the holographic image **originally required illumination** with a **laser light** similar to that used in its **manufacture** (either **reflected off** or **shone through** the **photographic film**). However, later advances produced the familiar **"rainbow" holograms** that can be viewed in **normal light**.

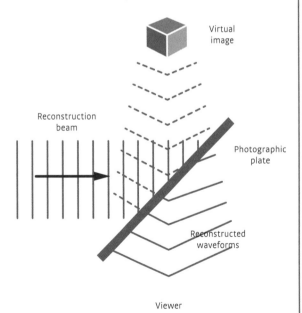

Virtual image

Reconstruction beam

Photographic plate

Reconstructed waveforms

Viewer

APPLICATIONS

- **3-D imaging**: Because the **light field** around the object is recorded from a **range of angles**, it's also possible to **view it from different angles in the hologram**.

- **Security**: Holograms with **complex embedded detail** can be used to make **forgery of banknotes**, **credit cards**, and other **important documents** more **difficult**.

- **Advanced optics**: Holograms can be used to **"record"** the **behaviour** of **complex optical systems**, allowing **images** or **light sources** to be **rapidly "processed"** through **interaction** with the hologram.

HOLOGRAPHIC DATA STORAGE

Holography **isn't limited to capturing 3-D information about real-world objects** – **interferometry** can be used to **store entirely different images** that are **visible from different angles**, and the images may simply be a **dense array of points** representing **binary 1s and 0s**, allowing holograms to **store digital data**. In a **thick, light-sensitive material**, different holograms can be **stored at different depths**, providing a **high-capacity alternative** to **traditional optical discs**.

INFRARED AND ULTRAVIOLET

Visible light covers a small range of wavelengths that our eyes have evolved to see. Infrared and ultraviolet are electromagnetic radiations that lie just beyond the limits of our vision.

INFRARED RADIATION

Wavelengths: 700 nanometres → 1 mm

Infrared radiation has **wavelengths longer than light**, and **correspondingly lower frequencies**). It is best thought of as **heat radiation**, **emitted by all objects** in the **Universe** except the **very coldest**, and **transferring energy** from the **warm object** to its **surrounding environment**:

 Near infrared: high-frequency, short wavelengths, emitted by **hot objects** including **stars**.

 Mid-infrared: lower-frequency, longer wavelength rays emitted by **warm everyday objects** including **humans** and **animals**.

 Far infrared: low-frequency, very long wavelength rays emitted by some of the **coolest materials** in the Universe including **interstellar dust**.

CHANCE DISCOVERY

Infrared was **discovered by accident** when astronomer **William Herschel** was attempting to **measure** the **temperatures** associated with **different colours of sunlight** split through a **spectrum** – he found the **temperature** of his **thermometer shot up** when it was placed in the area **just beyond the reddest light**.

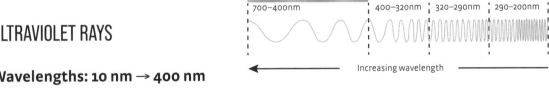

Visible light	UVA	UVB	UVC
700–400nm	400–320nm	320–290nm	290–200nm

← Increasing wavelength —

ULTRAVIOLET RAYS

Wavelengths: 10 nm → 400 nm

Ultraviolet (UV) has **higher frequencies** and **shorter wavelengths** than **visible light**. Wavelengths from **400 down to 100 nanometres (nm)** are **broadly divided** into the **near, middle,** and **far ultraviolet** (UVA, UVB, and UVC respectively), while wavelengths from **100 to 10 nm** are termed **Extreme Ultraviolet**.

UV radiation carries enough **energy** to trigger **chemical reactions** and **harm human cells**, but fortunately **Earth's atmosphere blocks** the **more harmful, shorter wavelength forms** from reaching the surface.

The **higher frequency** and **energy** of **UV** mean that it is **naturally emitted** by **surfaces hotter** than those that **shine in visible light**. The **Sun** produces substantial amounts of UV in line with its **black-body radiation curve**, and **hotter** and **more massive stars** can emit **most of their light in ultraviolet**.

RADIO AND MICROWAVES

Electromagnetic radiation ranging beyond the infrared forms the radio part of the spectrum. The shortest and highest-energy radio waves, with their own unique properties, are also called microwaves.

RADIO PROPERTIES

The **low energies** of most **radio waves** mean they can be produced by a **wide variety of different natural processes**, often involving **changing electromagnetic fields**.

Wavelengths: 1 mm → infinity

They have **little or no effect** on **objects they come into contact with**, but their **long wavelengths** make them **hard to disrupt**, passing **through** and **around** most **obstacles**. This makes them **ideal** for **sending signals** at the **speed of light**.

RADIO TECHNOLOGY

A **radio transmitter** is essentially a **long antenna** within which a **high-frequency alternating current** sends **electrons vibrating up and down**.

The **receiver** has a **tuneable "resonant" circuit** with an **antenna of its own**. **Electrons** will **vibrate** in the **antenna** when influenced by a **passing radio wave** of the **selected frequency**, generating a **current** that **mirrors** the **original signal** and can be used to **reconstruct sound waves**, **pictures**, or **digital data**.

MICROWAVES

Wavelengths: 1 mm → 1 m

These are the **highest-energy radio waves**, with **unique properties**:

• They **interact** more with **matter** – a **microwave oven**, for example, **traps microwaves** in an **isolated cavity** where they cause the **molecules in food** to **vibrate**, **heating** them up.
• They can be **transmitted** in **highly directional beams** but are **blocked by large obstacles** and can only **travel by line of sight**.

Microwaves are generated in devices called **magnetrons**, which cause **electrons** to follow **spiral paths** and **generate currents** that produce **oscillating electromagnetic disturbance** at the **right frequency**.

Hot cathode emits electrons that travel outwards

Stable magnetic field B

X-RAYS AND GAMMA RAYS

Beyond ultraviolet lie the highest-energy forms of radiation with the shortest frequencies and so much power that they can pass through most materials. These are the X-rays and gamma rays.

X-RAYS

Wavelengths: 0.01 nm → 10 nm

X-rays are generated by the **hottest materials** in the Universe – **gases heated to millions of degrees** in the **atmospheres** of **stars** and the **space between galaxies**. Their **penetrating radiation** has the **power to damage living things**, but our planet's **atmosphere blocks X-rays from space**, and **on Earth** they can **only be generated by artificial means** (often using **high-voltage vacuum tubes**)

X-ray **applications** include:

- **Radiography** – using the **penetrating power** of X-rays to investigate the **internal make-up** of **structures** and **living things** (the **denser the material**, the **more X-rays are absorbed** and the **fewer reach the photographic plate** or **detector** behind).
- **Crystallography** – fired into different solids, X-rays experience **diffraction** – the **spacing of internal molecules** acts as a **diffraction grating**, allowing the **structure** of substances such as **proteins** to be **analysed**.

GAMMA RAYS

Wavelengths: Less than 0.01 nm

Gamma rays are released during **radioactive decay** in the **atomic nucleus** (as a means for an **unstable nucleus** to **shed energy**), and also by **violent cosmic phenomena** such as **monster black holes** and **dying massive stars**. They **pass straight through most materials**, but are **absorbed** in the **thickness of Earth's atmosphere**.

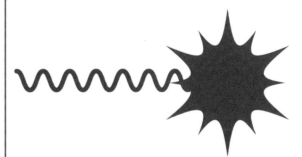

Ground-based gamma-ray observatories look for **showers of particles** released when a ray makes a **rare direct hit** on an **atmospheric gas molecule**.

Space-based telescopes use a "**coded mask**" of **dense material** that casts its **shadow** onto a **grid of gamma-ray detectors** and therefore reveals the **direction** of a **gamma ray's source**.

DISCOVERY

X-rays were discovered by **Wilhelm Röntgen** in 1895, during his experiments with an **electric vacuum tube** similar to the **cathode ray tube**. **High voltages** created **electrons** with **so much energy** that their **impacts** with **other parts of the tube** released **X-rays**, which **passed straight through barriers** and **"fogged" photographic plates** stored nearby.

WAVES AND PARTICLES

Although light and other forms of electromagnetic radiation usually behave as waves and experience the same phenomena as other types of wave, some aspects of their behaviour can only be explained by particle-like properties.

PACKETS OF LIGHT

Many processes related to **electromagnetic radiation** involve the **emission** or **absorption** of **small amounts of light** with a **single associated wavelength** and **frequency**. In 1905, **Albert Einstein** suggested these small units or "**quanta**" of light are not a mere **by-product** of the processes involved, but are an **inherent aspect** of **light itself** – light has **wavelike properties**, but it is **emitted in small packets**, now known as **photons**.

Einstein's idea helped kickstart the revolution of **quantum physics** (see p.127), but **what does it say about light itself?**

PHOTON FEATURES

Although photons are often depicted as small "**bursts**" of **waves**, **experimental evidence** suggests that they are best thought of as **single isolated waves** or "**solitons**". A **stream of light waves** that **appears to be continuous** is actually made of **photons following each other in close succession**.

A **single photon** consists of a **sinusoidal (sine-wave-shaped) disturbance** in **electric** and **magnetic fields**, **perpendicular** to **each other** and to the **direction of travel** (as in **Maxwell's model of electromagnetic waves**):

PHOTON ENERGY

A photon of **wavelength** λ carries a **small amount of energy** E, determined by

$$E = hc/\lambda$$

where c is the **speed of light in a vacuum** and h is **Planck's constant**.

In terms of **frequency** f, this can also be written $E = hf$, so photons of **higher frequency** deliver **more energy** (and it **takes more energy** to create a **photon of higher frequency**).

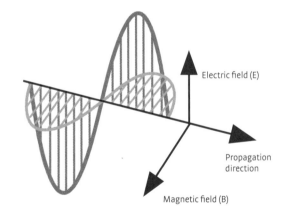

Electric field (E)

Propagation direction

Magnetic field (B)

In the **absence of environmental factors** to **drain their energy away** (for instance, in the **vacuum of empty space**) the **perpendicular electric and magnetic fields** act to constantly **regenerate** and **reinforce** each other, giving photons effectively **infinite range**.

FORMS OF ENERGY

Energy comes in many forms, and is fundamental to understanding every physical process in the Universe. The specialist field of physics devoted to the understanding of energy and its transfer is known as thermodynamics.

ENERGY AND TEMPERATURE

Physics deals with energy in many forms, ranging from the **potential energy** of **objects** held within **force fields** or **atoms** and **molecules** susceptible to a particular **chemical reaction**, to the **kinetic energy** of **objects in motion**, and the **electromagnetic energy** of **fluctuating fields**.

The temperature of an object is simply a **measure** of the **kinetic energy of its molecules** – a property that it **readily transfers to its surroundings** through various processes.

ABSOLUTE ZERO

As materials get **colder**, the **kinetic energy** of their particles **falls**. If **chilled** to a **sufficiently low temperature**, they will **lose their kinetic energy completely** as the **particles stop moving**. This temperature, **common to all materials in the Universe**, is known as **absolute zero**, the **coldest possible temperature** (equivalent to −273.15°C).

HEAT AND THERMODYNAMICS

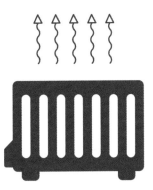

The mysterious property we think of as **heat** is **energy in motion between systems** (isolated or **self-contained objects** or **groups of objects**). The **study of heat** and its **motion** and **transfer** is known as **thermodynamics**.

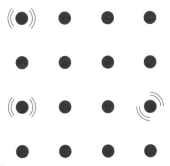

Whether we think of a material as **hot** or **cold** is a **measure of its temperature** (the **kinetic energy of its particles**) but also of its **tendency to radiate or absorb heat**.

Physicists **define heat very precisely** as **energy transferred**:
- **without matter itself being moved in bulk** – because **such movement does not transfer heat energy** but rather **moves matter of a certain temperature from one place to another**;
- **without the energy doing mechanical work** – because **such work "uses up" the energy**, changing it to **other forms** so that it **cannot be used to heat other systems**.

TEMPERATURE MEASUREMENT

A material's temperature is a reflection of the average kinetic energy of its particles.
Various techniques can be used to measure it, and various scales used to record it.

MEASURING TEMPERATURE

In order to **measure temperature**, scientists use **materials** and **instruments** with a **large-scale "macroscopic" property** that **changes** as a **reflection** of the **kinetic energy of their constituent particles**.

Ideally the change should be **linear** (or at least **mathematically simple**) and **obey** the **same rules** over a **wide range of temperatures**.

For example in a **mercury thermometer**:

- **Liquid metal** contained in a bulb **expands at a uniform rate** between the **freezing** and **boiling points** of **water** (due to its **increasing kinetic energy**).
- **Contact** between the **thermometer bulb** and its **surroundings warms** or **cools** the **liquid inside**.
- The **expansion** or **contraction** governs the **height** of a **thin column of mercury** fed from the bulb.

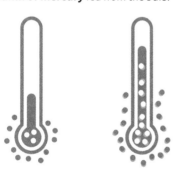

TEMPERATURE SCALES

Temperature scales are constructed by **recording the properties** of a **measuring device** at **opposite ends** of a **useful temperature range** and then **splitting up the range** between them on a **graduated scale**.

Common temperature scales:

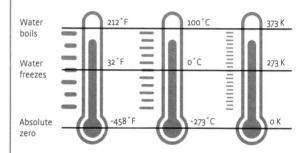

Water boils	212°F	100°C	373 K
Water freezes	32°F	0°C	273 K
Absolute zero	-458°F	-273°C	0 K

- **Fahrenheit**: Sets 32°F as the **melting point of water** and 212°F as its **boiling point**, dividing the **intervening space into 180°**. 0°F is the **freezing point** of a mix of equal parts **water**, **ice**, and **salt**.
- **Celsius**: Sets the **freezing point of water** as 0°C and the **boiling point** as 100°C, with **100 degrees (intervals)** between them.*
- **Kelvin**: Uses **identical intervals to Celsius**, but sets 0 K as **absolute zero**.* On the Kelvin scale, the **freezing point of water** is 273.15 K.

* Informal definition – the official definitions of these scales are far more rigorous.

ALTERNATIVE THERMOMETERS

While most thermometers rely on the **physical expansion of materials with rising temperature**, there are other approaches:

- **Electronic thermometers** measure **changes** in the **ability of certain materials** to **conduct** or **resist electricity**.
- **Pyrometers** and **bolometers** collect the **radiation emitted from an object** in order to measure its temperature.

HEAT TRANSFER

Heat can be moved through and between materials and objects using three distinctly different mechanisms, known as convection, conduction, and radiation.

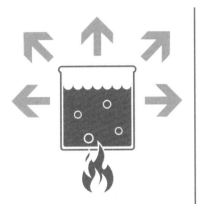

CONVECTION

Convection is a **bulk movement** of **hot material** through **cooler surroundings**.

- Because **cooler** materials tend to be **denser** and **hot** ones **less dense**, warmer materials will (in a **uniform material**) **rise through colder surroundings** until they find a region of **equal density** and **temperature**.
- If a material is **heated from below, convection cells** will form as **cool material** that **moves in** to **replace** the **rising warmer material** is **itself heated**.

CONDUCTION

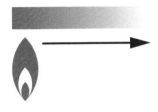

Conduction is the **motion of heat** through the **direct transfer of kinetic energy** between **particles** on a **microscopic scale**.

- **Hot**, **fast-moving** particles **collide** with **cooler, slower-moving** ones, **transferring** some of their **kinetic energy** and **increasing** the **motion** and **temperature** of the **cooler particles**.
- However, the particles stay in the **same relative locations** and **do not move great distances** through the heat-conducting medium.
- Because the **hotter** particles **lose some kinetic energy** in the process, **temperatures across the conductor** tend to **even out** – unless a **heat source** of some sort **continuously boosts** the **kinetic energy** of particles at **one end of the conductor**.

RADIATION

Radiation is the **transfer of heat** through **electromagnetic rays**.

- **All objects** and **materials** made of **normal matter** are **constantly absorbing radiation** from their surroundings, which has a **heating effect**, and **emitting thermal radiation** into their environment (see p.61).

- The **wavelength** and **energy** of **emitted radiation** changes with the **temperature** of the **emitting material**, but for most objects it is **concentrated** in the **infrared region** of the **spectrum**.

- Unlike the other forms of **heat transfer, radiation can take place across a vacuum**, such as **space**.

CONDUCTING ELECTRICITY AND HEAT
Metals conduct electricity and also tend to be **good conductors of heat**. Heat conduction is partly linked to their **lattice-like structure of fixed atoms** but also to the **mobile electrons** that can **move through the lattice**, and **transfer energy** as well as **electricity**.

ENTROPY

In the science of thermodynamics, two related properties called entropy and enthalpy describe the way that energy changes its distribution within and between thermodynamic systems.

ENTHALPY

The **enthalpy** of a system, denoted H, is its ***total*** energy. Enthalpy **combines** many **different** and **distinct forms of energy**:

- **Kinetic energy** of particles.
- **Potential energy** of various kinds.
- Energy required to **create the system**.
- Energy needed to **displace the environment around it**.

It is defined by the equation $H = U + pV$, where U is the system's **internal energy**, p its **pressure**, and V its **volume**.

Enthalpy is always **relative**, so it is **impossible to measure absolutely**. Instead, thermodynamics deals with *changes* to enthalpy:

- When a process **adds energy** to a system, it is **endothermic**.
- When a process **removes energy** from the system, it is **exothermic**.

UNAVAILABLE ENERGY

In theory, one could devise a **theoretically perfect mechanism** to **harness differences** in the **kinetic** or **thermal energy** of particles and **extract mechanical work**.

But it would be **impossible to extract** *all* **the energy from a system**. A **certain amount** of **thermal energy** always remains **unavailable to do work**. This is the system's **entropy** (denoted S).

Entropy is often interpreted as a **measure of the disorder within a system on a microscopic scale** – the degree to which **energy is scattered evenly** between the particles and so **cannot be used to do work**.

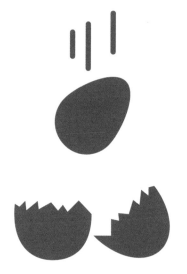

FREE ENERGY

Two equations describe the **energy available** to do **work** in different situations:

- **Helmholtz free energy** F, the **work obtainable** from a system held at **constant temperature** and **volume**.

If T is the system's **absolute temperature** in **kelvin**, then

$$F = U - TS$$

- **Gibbs free energy** or **free enthalpy** G, the **maximum work obtainable** from a system at **constant pressure** p, and **temperature** T **without changing the volume** V:

$$G = U + pV - TS$$

$$\text{or}$$

$$G = H - TS$$

LAWS OF THERMODYNAMICS

Four laws govern the thermodynamic behaviour of systems – they turn out to have far-reaching consequences for the whole of physics and even the future of the Universe itself.

FOUR LAWS

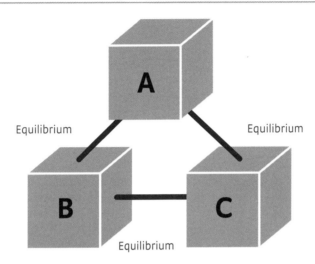

Equilibrium   Equilibrium

Equilibrium

Zeroth law: If **two objects** are in **thermal equilibrium** with a **third**, then they are also in **equilibrium with each other**. This is the **most essential law**, but was **discovered after the others**, hence its place, and unusual name.

First law: **Heat** and **work** are both forms of **energy transfer**. The **internal energy** of a **closed system** changes if **heat is transferred outwards** (by the **system doing work**) or **inwards** (by **work done to the system**).

This law demonstrates that the fantasy of a **perpetual motion machine** is **impossible**, since **extracting any work** from a system inevitably leads to a **fall in its energy**.

Second law: The **entropy of a closed system never decreases**, and a system will **evolve towards thermal equilibrium unless external work is done** (and **energy expended**) to **prevent it**.

This law is often encapsulated in the phrase **entropy increases** – it also provides a **directionality** or "**arrow of time**" to **physical laws** and explains why, for example, a **dropped glass** will **shatter into pieces**, but **dropping the fragments will never see them reassemble into their original form**.

Third law: The **entropy** of a system approaches a **constant value** only as the **temperature approaches absolute zero**. In other words, as the **kinetic energy** of all the particles present **falls to zero, disorder disappears**.

The **minimum constant value of entropy** in this situation varies – in **crystals** and other **orderly structures** it is itself **zero**, but **disorderly materials** such as **glass** may retain some **residual entropy** even at **absolute zero**.

THE FATE OF THE UNIVERSE

The **second and third laws together** determine the **likely fate of the Universe** (assuming no other forces intervene). Since there is **no way of providing energy from outside** to **reverse it**, **entropy inevitably increases**. This dooms the Universe to a **"heat-death" of slow cooling** as the **temperature of its particles averages out** and **falls towards absolute zero**.

HEAT CAPACITY

Apply the same amount of heat to two different materials, and their temperatures will rise by different amounts. This is due to differences in a property known as heat capacity.

HEATING MATERIALS

Different materials **respond to heat** in different ways depending on their **internal structures**, and this affects **how much of the thermal energy** supplied to a material **ends up in the form** of **translational kinetic energy** that is **measured as temperature**.

The **more complex** the material's **internal structure**, the **more "degrees of freedom"** it has and the **more places** there are for the **thermal energy to go**:

Monatomic gases such as **helium** are made of **free-floating atoms**, so when they are **supplied with thermal energy** it **goes directly into kinetic motion** and they **heat up rapidly**.

In **more complex diatomic gases** such as **chlorine**, **thermal energy** can also be **absorbed by interatomic bonds**, causing them to **vibrate**, or **rotate**. As a result, **supplying the same amount of energy** to the **same number of particles** produces a **smaller temperature increase**.

Liquids and **solids** with **more complex structures** respond to thermal energy in different ways. **Metallic solids** tend to **heat up fairly rapidly** since, **despite the number of bonds**, they are **tightly constrained**.

Water absorbs a lot of energy for a **small change in temperature** because its **relatively complex molecules** and **wide range of intermolecular bonds** give it **many different degrees of freedom**.

MEASURING HEAT CAPACITY

A material's **response to heating** is described by its **heat capacity**. For ease of comparison, **two measurements** are widely used:

- **Specific heat capacity**: the **amount of heat in joules** that is **required to raise** the **temperature** of **1 kg of material** by **1 kelvin**.
- **Molar heat capacity**: the **amount of heat required** to **raise 1 mole*** of a material's **constituent atoms** or **molecules** by **1 kelvin**.

Specific heat capacity has most uses in **engineering** and **everyday life**, but **molar heat capacity** is more useful for understanding the **response to heat** of a **material's individual particles**.

Substance	Specific h.c. (J/kg/K)	Molar h.c. (J/mol/K)
Water	4181	75.3
Iron	412	25.1
Aluminium	897	24.2
Gold	129	25.4
Wood	2380	N/A

* 1 mole $\approx 6.022 \times 10^{23}$ – the number of particles required to equal a single particle's atomic mass in grams.

CHANGES OF STATE

While the addition or removal of heat energy from an object or system normally causes an increase or decrease in its temperature, that's not always the case – sometimes the energy is diverted to, or derived from, the breaking, or making, of chemical bonds.

PHASE TRANSITIONS

When a material **changes its state** (the **large-scale arrangement** of its **atoms** or **molecules** – most commonly, whether it is a **solid**, **liquid**, or **gas**) it is said to go through a **phase transition**.

Phase transitions occur when the **environment** and **conditions affecting a material** have **changed** in a way that **no longer supports the existing state** but instead is **more suited to a different state** – for instance, when a **liquid's**

temperature rises to the point where it would **normally be found as a gas**, or **falls** towards the point where it would **normally be solid**.

Transition involves a **large-scale reorganization** of **bonds** between **atoms** or **molecules**. **Breaking bonds** typically requires the **addition of energy** and is an **endothermic process**, while **making bonds releases energy** and is **exothermic**.

As a result, phase transitions involve the **absorption** or **release** of a **considerable amount of thermal energy** with **no accompanying change in temperature**. **Endothermic transitions** lead to an **increase in entropy** (**breaking the bonds** is a form of **thermodynamic work**) while **exothermic transitions** lead to a **decrease in entropy**.

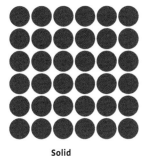

Solid

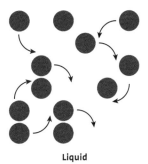

Liquid

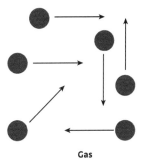

Gas

LATENT HEAT

The **energy absorbed** or **released** in a **phase transition** is known as the process's **latent heat** or **enthalpy**. Measured in **joules per kilogram** (**specific latent heat**) or **per mole** (**molar latent heat**), it is the **energy released** or **absorbed** by the **forging** or **breaking of bonds** across a set amount of the material.

Latent heats for water	Specific (kJ/kg)	Molar (kJ/mol)
Latent heat of vaporization (boiling):	2,264	40.7
Latent heat of condensation:	-2264	-40.7
Latent heat of fusion (melting):	334	6.0
Latent heat of solidification:	334	6.0

HEAT ENGINES

Thermodynamics developed from a desire to understand the physical principles behind the operation of steam engines, and thermodynamic processes can still be usefully understood in terms of the operation of more generic "heat engines".

HEAT CYCLES

As first defined by **Nicolas Sadi Carnot** in the 1820s, a heat engine is **any thermodynamic system** that **converts heat energy into mechanical work**. Most examples make use of the **changing temperature** of a **working fluid** in a **repeating "heat cycle":**

- **Working fluid** (liquid or gas) is **heated**.
- **Energy** from fluid is **harnessed** to do **work**.
- **Excess heat** is **drained** into **low-temperature "sink"**.
- **Working fluid** returns to **original temperature** and is **reheated**.

Work can be **extracted** from the **working fluid** after **heating** by means of **physical principles** such as the **gas laws** (see p.49). For example, **changing** the **temperature of a gas** within a **closed volume increases** or **reduces** the **pressure** it **exerts** on its **container**.

A **heat pump** is the **opposite of a heat engine** – a device that uses **mechanical forces** to **transport heat from one part of a system to another**. **Fridges** and **air-conditioners** both work on this principle, using a **working fluid** that **draws heat** from one part of the system and **releases it** elsewhere.

No heat engine or **pump** can be **100 per cent efficient** – heat is **inevitably lost** into the **components** of the **pump** and the **wider environment**, **increasing** the system's **entropy**.

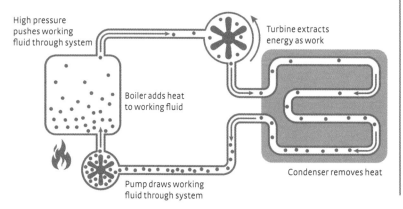

High pressure pushes working fluid through system

Turbine extracts energy as work

Boiler adds heat to working fluid

Condenser removes heat

Pump draws working fluid through system

ENGINE TYPES

Steam engines make use of the dramatic **change in pressure** exerted when **liquid water** is **converted to vapour**, using it to **push a piston outwards**. **Sudden cooling** and **condensation** then create a **drop in pressure** that **draws the piston back in**.

In an **internal combustion engine**, **fuel** is **vaporized** before being **injected** and **ignited**. This creates a **less dramatic pressure/volume change** and allows for a **more compact engine**.

STATIC ELECTRICITY

Electric charge is a property of many subatomic particles, but large-scale matter is usually electrically neutral. However, hidden electric charge can manifest itself in nature as static electricity.

CHARGE BUILD-UP

Static electricity is a **build-up** of **non-moving electrically charged particles** on **surfaces**. It occurs when something causes a **transfer of electrons** (the **negatively charged particles** in the **outer layers of atoms**) **between** or **within materials**.

The material that **gains electrons** is said to have **negative polarity**, while the one that **loses them** gains a **positive polarity** due to the **unbalanced positive charge** remaining in the **centre** of its **atoms**.

Examples of this **transfer** include:

Certain materials **rubbing against each other** (for instance, **amber** and **wool**, **glass** and **silk**, and **rubber** and **human or animal hair**).

Convection of **ice particles** within **storm clouds** (creating a **net positive charge** at the **top** of the cloud and a **negative charge** at the **base**).

STATIC DISCHARGE

If **charge** is **not able to flow away**, the **electric field** between **areas of opposing charges** can become **so strong** that it affects the **intervening air**. **Electrons** are **stripped away from molecules**, **ionizing** them and turning the **air** into an **electrical conductor**, which allows **current** to **flow** and **equalize** between the **surface**.

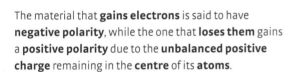

- **Lightning** is a **natural form** of **static discharge** – it occurs **within clouds** and **between** their **bases** and the **ground**.
- **Pointed objects concentrate charge** while **smooth surfaces spread it out** and allow **greater differences** to **build up** – **lightning conductors** offer a **path of least resistance** that **attracts lightning**.

ELECTROSTATIC INDUCTION

Because both **moving electric current** and **static charges** create **electric fields**, they can **affect** other **nearby materials without direct contact**. For instance, the **negative charge** at the **base** of **storm clouds** induces a **build-up** of **positive charge** in the **ground below**.

- A **Van de Graaf generator** creates an **artificial build-up** of **electric charge** on **separated metal spheres** using a **rotating rubber band**.

CURRENT ELECTRICITY

Electric current is the flow of charged particles from one place to another through a material known as a conductor. It is the form of electricity we can most easily put to work.

CONDUCTORS

The **flow of electric current** involves the **movement of charge** through a material called a **conductor**. In practice, current is usually a flow of **negatively charged electrons** (in a **molten material** it can also be a flow of **positively charged ions**).

Particles with similar charges repel each other, while **opposites attract**, so electrons have a tendency to **flow from areas of negative polarity** to those with **positive polarity**.

Metals are the **best conductors**, since their **bonding involves atoms releasing electrons** from their **outer layers** into a **diffuse "sea"** that can be made to **flow through** and **around** the **solid lattice structure**.

CURRENT CHARGE AND VOLTAGE

The **strength** of a current (denoted I) is **measured** in terms of the **amount of electric charge** (q) passing through a **cross-section of conductor each second**. Because the **elementary unit of charge** (that of a **single electron**, denoted q_e) is so **tiny**, scientists and engineers habitually use a **much larger unit**, known as the **coulomb**:

$$1 \text{ coulomb} = 6.24 \times 10^{19} \ q_e$$

The **basic unit of current**, known as the **ampere** or **amp**, is therefore:

$$1 \text{ amp} = \text{current flow of 1 coulomb} \\ \text{per second}$$

Electric fields (such as those generated between areas of **different polarity**) create a difference in the **electrical potential energy** of particles in **different locations**, which is released as **electrical energy** when current flows. This **potential difference** (or **voltage**) V is measured in **volts**:

$$1 \text{ volt} = 1 \text{ joule of energy} \\ \text{per coulomb of charge}$$

Potential difference is defined so that **positive "conventional current" always flows from higher voltages to lower ones**.

CONFUSING CONVENTION

Despite most electric currents involving movement of **negatively charged electrons**, the long-standing convention is to **depict current** as a **movement of positive charge**.

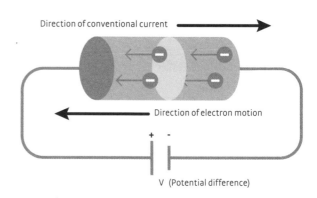

Direction of conventional current →

Direction of electron motion

V (Potential difference)

CIRCUITS

Electric current will only flow through a conductor when it is formed into a closed loop and driven by a potential difference. Different circuit arrangements can behave in different ways.

ELECTROMOTIVE FORCE

Current **needs somewhere to go** – it does not normally flow through **isolated materials** because the movement of charge will rapidly **change the polarity between different locations**, producing a **potential difference** that tends to **counteract the flow**.

An **electrical circuit links together various conducting components**, but in order to create a **constant flow** of current, it's still necessary to have an influence that **pushes charge carriers around**.

This driving influence is known as the **electromotive force** (EMF) – though in practice it's **not a force measured in newtons**, but instead a **potential difference measured in volts**.

EMF SOURCES
Cells: **chemical reactions** cause **current to flow** through a **liquid** or **semi-liquid** material, creating **potential difference between one end and the other**. **Solid conductors**, known as **electrodes**, act as either **emitters** of electrons (the **cathode** or **negative electrode**) or **absorbers** of them (the **anode** or **positive electrode**). A battery is technically a **series of individual cells**.

Thermoelectric and photoelectric cells: **energy** from **heat** or **light** powers flow of current and **generates potential difference**.

Dynamo: a **changing magnetic field** drives the **movement of charge carriers** through **induction** (see p.100).

CIRCUIT LAYOUTS

Once a current is flowing, it can be **driven through each of a series of components in turn** (a **series circuit**) or **split apart** so that it **passes along two or more routes simultaneously** (a **parallel circuit**).

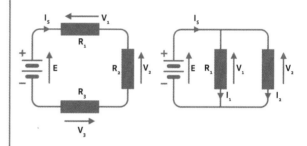

Series circuit:
- Current I remains **constant** as it **passes through all components in turn**, but **stops completely if any component fails**.

- **Loss of EMF** as current **passes through each component** (known as **resistance**) produces a **voltage drop** in each, $V_1, V_2, V_3 \ldots$

- The **total voltage drop** ($V_1 + V_2 + V_3 \ldots$) is **balanced** by the **supplied EMF** E, so the current is **refreshed** and **keeps flowing**.

Parallel circuit:
- Current I is **split between branches** into $I_1, I_2 \ldots$ A **break in one branch** simply **sends more current through those that remain intact**.

- **Less current** flows along branches whose components have **greater resistance**.

- **Voltage drop** V across each path is the **same**, and **identical** to the **supplied EMF** E.

ELECTRICAL COMPONENTS

An electrical circuit can contain several different components, each of which produces an effect known as resistance that slows and diminishes the flow of current.

WHAT IS RESISTANCE?

Resistance is simply a **material's opposition** to the **flow of electric current**. As **charge carriers move through any conductor**, they inevitably **interact** with their **surroundings** (and **each other**) in ways that **slow them down** and **shed energy** into their surroundings – similarly to **friction** in **mechanics**.

Resistance (denoted R) is determined by **Ohm's Law**: $R = V/I$, where I is the **current running through a component**, and V is the **voltage drop** from one side to the other due to this **loss of impetus**.

Resistance is measured in **ohms**: 1 ohm = 1 volt per coulomb. Materials with very **high resistance** are termed **insulators**.

Conductance (denoted G) is the **inverse of resistance** – a lesser used measure of **how well a material conducts**, rather than **how much it resists**.

COMPONENTS

A component is simply an **electrical device** with a **specific task** in a **circuit**. Examples include:

Resistors: deliberately placed to **adjust** the **flow of current** or the **voltage difference** in **one part of a circuit**, or simply to **dissipate energy as heat** (**electric heaters** and **incandescent lightbulbs** both use **high-resistance materials** that heat them to **high temperatures**: the **power** dissipated in **watts** is given by $P = VI$).

Switches: **block** or **allow** the **flow of current** depending on their **state** (which may be **manually changed** or **controlled by an external factor** such as a **time clock, thermostat, light sensor**, etc.).

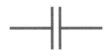

Capacitors: **store electrical energy** as **potential difference** by accumulating **opposing static charges** on two **conducting plates** separated by an **insulated gap**. The capacitor is **discharged** by **bridging the gap**.

Motors: **convert** the **energy** of **flowing electric currents** into **mechanical motion**.

Diodes: permit **electricity** to **flow in only one direction**.

Inductors: **store electrical energy** in a **magnetic field**.

MAGNETISM

The familiar effects of magnetism are just one aspect of the electromagnetic force – a fundamental force of nature. Magnetism and magnetic fields are in fact the results of electric charges in motion.

MAGNETIC FIELDS

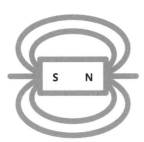

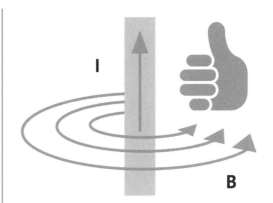

A magnetic field has **many similar properties to an electric one**:

- Two **opposing polarities** (north and south).
- **Attractive forces** between **opposite poles** and **repulsive forces** between **similar poles**.
- An ability to **magnetize other susceptible objects**, exerting an **attractive force** on them and even **transferring permanent magnetism**.

Individual moving **electric charges** and **electric currents** both **generate magnetic fields** that exert **force** on a **line tangent to a circle** around the charge's **direction of motion**. The **strength** of a magnetic field (in terms of **magnetic flux density** B, and measured in **teslas**) is given by

$$B = \frac{\mu_0 I}{2\pi r}$$

where r is the **distance** from a **long straight wire** carrying a **current** I. The **direction** of the field (from **south** to **north**) is shown by the simple **right-hand grip rule**: with the thumb pointing in the direction of the **conventional current**, the curl of the other fingers follows the direction of the **field**.

MAGNETIC PERMEABILITY

Known as the **permeability of free space** or simply the **magnetic constant**, **physical constant** μ_0 defines the **relation of magnetic force** to **space** and **electric charge**. Its value is 1.2566 × 10⁻⁶ (measured in **henrys per metre**).

Another constant, the **permittivity of free space** ε_0, defines the **strength of electric fields** in a similar way – the fact that

$$\varepsilon_0 \mu_0 = 1/c^2$$

(where c is the **speed of light**) shows the **fundamental relationship** between **electricity**, **magnetism**, and **light** (now understood to be a **moving electromagnetic field itself**).

FIELD COMPARISONS

Average strength of **Earth's magnetic field**: 50,000 nanoteslas (nT).
Typical **fridge magnet**: 10 million nT.
Large sunspot (concentration of magnetic field on the Sun): 0.3 T.

MAGNETIC MATERIALS

The ability of bulk materials to respond to or even permanently hold a magnetic field depends on several aspects of their internal structure, and particularly on the motion of tiny electric charges within them.

MAGNETIC MOMENTS

Magnetism and magnetic fields are a **consequence of electric charge in motion** – "magnetic charge" is **not a fundamental property of particles in the same way as electric charge**.

Instead, the magnetic properties of materials are generated by the **flow of tiny electric currents** on the **atomic** and **subatomic scale**:

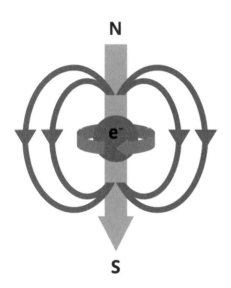

- All **subatomic matter particles** have a property of **spin** (analogous to **angular momentum** in **mechanics**).
- The **action of spin** on **electrically charged particles** (particularly **electrons**) produces the effect of a **tiny loop of circulating electrical current**.
- The spinning charge therefore gives rise to a **tiny magnetic field** known as the particle's **magnetic moment**.

TYPES OF MAGNETISM

In most bulk materials, **internal magnetic moments** are **arranged at random** and tend to **cancel out**. However, in the **right circumstances**, the moments can **align** to create **large-scale magnetic effects**. These include:

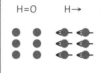

- **Diamagnetism**: The **polarity** of individual particles **aligns** to **oppose an external magnetic field**, producing a **weak repulsive force** between **field** and **object**.

- **Paramagnetism**: If **atoms** in a material contain **uneven numbers of electrons** in their **outer structural shells** (see p.112), the "unpaired" electrons will **line up** when an **external magnetic field** is applied, generating their **own field** in response and **temporarily magnetizing** the material. The **internal magnetism disappears** and the **moments usually disappear** when the **external magnetic field** is **removed**.

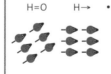

- **Ferromagnetism**: Certain **metals** (such as **iron**, **nickel** and **cobalt**) have **unpaired electrons** that show a **natural tendency to remain aligned** even after the influence of an **external field** is removed, producing **semi-permanent magnetism. Exposure to other magnetic fields** or **heating above the material's Curie temperature** causes the magnetism to dissipate.

AMPERE'S AND COULOMB'S LAWS

Particles with similar electric charges repel each other while those with opposite charges attract. Two important laws describe the force exerted between static charges and moving currents in these situations.

COULOMB'S LAW

This law describes the **attractive** or **repulsive force** between **two bodies**, each carrying an **electric charge** (for instance that built up from **static electricity**).

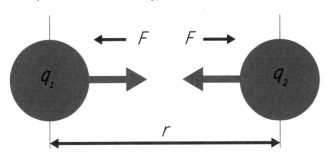

Coulomb's law takes the simple form

$$F = k_e \frac{q_1 q_2}{r^2}$$

where F is the **strength of the *repulsive* force between the charges**, q_1 and q_2 are the **charges themselves**, r is the **distance between them**, and k_e is **Coulomb's constant**, a reflection of the **ability of the medium around the charges to support an electric field**. When the signs (**polarities**) of the two charges are **similar**, the **resulting force** is **positive (repulsive)**, but when they are **opposite**, the force is **negative (attractive)**.

Note the similarity to **Newton's Law of Universal Gravitation** – the **force** between the objects **increases in proportion to the charge on each**, and **decreases in proportion to the square of the distance between them** as the **magnetic effect that underlies the force** becomes more **widely dispersed through space**.

Coulomb's law explains how objects that develop **static charge** (for instance, expanded **polystyrene packing chips**) are **attracted to others**, and why your **hair may stand on end** shortly before a **thunderstorm**.

AMPERE'S LAW

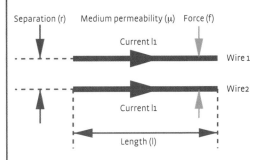

Ampere's law describes the **attractive force** between **two parallel wires carrying electric current**. The cause is the same – **interaction between the magnetic fields generated around the moving charge carriers** – but the resulting equation is distinctly different:

$$F = 2k_A \frac{I_1 I_2}{r}$$

Here the force is **proportional** to the **strength of the currents** I_1 and I_2 in **each wire**, and **inversely proportional** to the **distance between them** (rather than its square). The constant k_A, meanwhile, is the **magnetic force constant**, 2×10^{-7} newtons per ampere squared.

ELECTROMAGNETIC INDUCTION

Just as moving charges and currents can generate magnetic fields that exert physical force on other conductors, they can also cause current to flow inside them – a hugely useful phenomenon known as induction.

HOW INDUCTION WORKS

Strictly speaking, **electromagnetic induction** is the production of an **electromotive force** (**voltage difference**) in a **conductor**, rather than necessarily the **flow of current** – though the **two usually go together**.

The **EMF** is created by the influence of an **external magnetic field** (generated by a **magnetic material** or a **nearby electric current**) on the **magnetic moments** of **individual electrons** in the **conductor**. Its **strength** is described by **Faraday's law of induction**:

$$\varepsilon = -\frac{d\Phi_B}{dt}$$

The EMF ε around a **closed conducting path** is **equal** and **opposite to** the **rate of change of magnetic flux** Φ_B enclosed by the path.

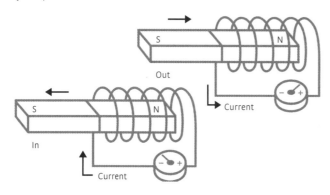

SHORT-LIVED EMF

Induction **only occurs** when the **magnetic flux around a conductor is changing**, so **direct-current circuits** produce only **brief bursts** of induction in **nearby conductors** when they are **switched on or off** – as the **magnetic field** around them **stabilizes, the induced EMF disappears. Constantly changing alternating current** offers a means of **sustaining induction**.

TRANSFORMERS

The transformer is a common electrical component that relies on **induction**. A simple form consists of a **square core of ferromagnetic metal**, with **wire coils** on **opposite sides** connecting it to a **primary** and a **secondary circuit**.

Current in the **primary circuit** creates an **induced magnetic field** in the **iron core**, which in turn creates an **EMF** and a **secondary current** in the **unpowered circuit**.

The transformer's usefulness lies in the fact that the **number of turns** N in the **primary** and **secondary coils** determines the **voltage** and **current ratios** between the circuits:

$$N_p/N_s = V_p/V_s = I_s/I_p$$

Note that the **power of each circuit remains the same**, because $V_p \times I_p = V_s \times I_s$.

However, once again, **induction only works when the current in the primary coil is changing**.

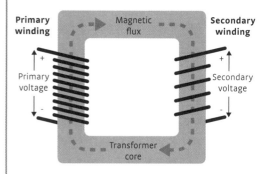

ALTERNATING CURRENT

Early experiments with induction found the phenomenon only worked with changing, rather than constant, current. To overcome this problem, engineers developed a system in which the current is constantly changing.

WHAT IS ALTERNATING CURRENT?

An alternating current (AC) is one in which the **direction of current** is **constantly changing** in a **repeating high-frequency cycle**. Perhaps surprisingly, this has **little effect on the possibility of harnessing it for electrical energy**.

AC is usually depicted as a **sine wave** in which both **voltage** V and **current** I **vary smoothly** between **positive** and **negative** as the **potential difference** and **EMF** vary in **strength** and **direction**, and **current reverses** as a result.

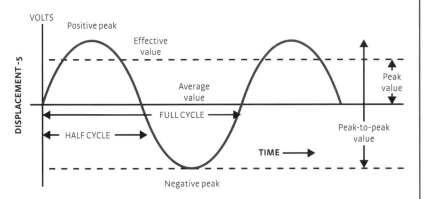

Because **current** and **voltage** are **constantly changing**, so is the **power** carried in an AC circuit. To simplify matters, **AC power** is **usually calculated** as an **average**, related to the **"root mean squared"** voltage:

$$\text{Average power } P = V_{\text{rms}}{}^2/R.$$

For a sine wave with peak voltage at V_{pk}, $V_{\text{rms}} = V_{\text{pk}}/\sqrt{2}$.

AC ADVANTAGES

- AC enables **electricity** generated with a **certain current** and **voltage** to **pass through transformers** that **work continuously** to **raise voltage** and **lower current**. Power losses to resistance are **much higher** when **currents are larger**, so **electricity transmission** is **more efficient** with a **high voltage** and **very low current**. Transformers **closer to the consumer** can then **adjust** the electricity to a **higher current** and **lower voltage** for **practical use** and **safety**.
- **Turbines**, **dynamos**, and many other **generators** also **naturally produce AC current**, which otherwise needs to be **"rectified"** in order to use it with some **DC components**.

THREE-PHASE AC

Most **large-scale power transmission systems** make use of **three-phase AC**, in which the **current is transmitted along three separate wires** that are each one third of a wavelength **out of step** with the others. This arrangement allows **more power** to be delivered, and ensures that the **actual power** being delivered is **roughly constant** at all times.

THE CURRENT WAR

Between 1880 and 1893, a fierce commercial battle was fought between **Thomas Edison**, who advocated **direct current**, and **George Westinghouse**, whose company had built the **first AC system**. **AC's victory** was eventually secured thanks to its **easier long-distance transmission** and the reliable **three-phase induction motor** invented by **Nikola Tesla**.

ELECTRIC MOTORS

An electric motor is a device that harnesses the force generated between electrical conductors and magnetic materials to convert electrical power into mechanical motion.

ROTARY MOTION

A motor works on the simple principle of **magnetic repulsion** between **fields of similar polarities**. However, most are designed to produce **circular** or **"rotary" motion** – for instance, driving the **rotation** of a **central axle**.

A **rotary motor** has two key elements:

- **Stator**: produces **magnetic fields** to drive the motion – usually remains **static**.
- **Rotor**: **spins on an axis** to generate the mechanical motion.

One element has **permanent magnets** (or **electromagnets**) fixed to it, while the other is wrapped with **tight coils** of **conducting wire** called **windings**. Current in the windings creates a **magnetic field** that drives **repulsion** from the **permanently magnetized element**, **forcing the rotor to turn** by a small amount.

ELECTROMAGNETS

Rather than use **permanent magnets** on one element, some motors use **electromagnets**. These take advantage of the natural properties of **ferromagnetism** to generate **particularly strong magnetic fields**. They also have the advantage that they can be **switched off and on**.

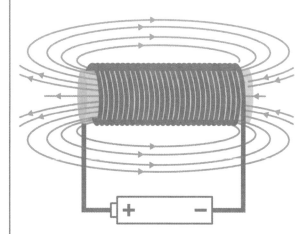

- **Conducting coil** around **metal core**.
- **Current flows through the coil** and generates a **weak magnetic field**.
- **Weak field** causes **magnetic moments** in the iron to **line up**, generating a much **stronger field**.

FLEMING'S LEFT-HAND RULE
Use this handy rule to calculate the **direction of thrust** generated by **current flowing through a magnetic field**.

- Middle finger: current.
- Index finger: field.
- Thumb points in the direction of thrust.

GENERATORS

Current electricity can be made in many different ways, but nearly all large-scale generators use the same basic principle, employing mechanical motion to power electromagnetic induction.

ELECTRICITY FROM MOVEMENT

Most generators **reverse the principle of the electric motor** – instead of using **changing electric currents** to drive **rotation**, they use a **rotating magnetic field** to produce **electric current**.

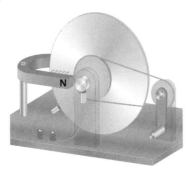

Early **dynamo generators** had **permanent magnets** mounted on the interior of a **fixed casing**. A **freely rotating armature wrapped in wire coils spun on its axis within the casing**, driven by an external force. As the wire coils encountered **constantly changing magnetic fields, induction produced current** within them.

The changing relationship between **magnets** and **wire armature** causes the current to **reverse** with each **half-rotation**, so in order to maintain **direct current**, the armature has a **commutator** so that the **connections between it** and the **rest of the circuit** are **also reversed at each half-turn**.

The success of AC means that **dynamo generators** are **rarely used these days** – instead, power is generated by **alternators**, in which the **magnets** are mounted on the **spinning core** while the **coils** are mounted in the **static casing**.

POWER SOURCES

The **rotation needed to drive a generator** can come from **many different sources**:

- Spinning **wind turbines**.
- Falling **water** in **hydroelectric dams**.
- **Tidal** energy.
- **Steam** produced in **nuclear** or **conventional power stations**.

FLEMING'S RIGHT-HAND RULE

Using this counterpart to the **left-hand rule** for motors reveals the **direction of current** generated when a **conductor moves through a magnetic field**:
- Thumb: direction of motion.
- Index finger: field.
- Middle finger points in the direction of the conventional current.

PIEZOELECTRIC EFFECT

The natural phenomenon known as piezoelectricity involves the production of electric current when certain substances are subjected to pressure. It lies at the heart of several modern technologies.

ELECTRICITY FROM PRESSURE

While a piezoelectric material appears **electrically neutral** from the **outside**, it has an **uneven distribution of charge within**, generating what is known as an **electric dipole moment**. This may be due to **internal concentrations of electrons in its molecules**, or the **shape of its repeating crystalline lattice**.

When **mechanical stress** is applied to the material, the **internal structure is rearranged on a microscopic scale**, leading to a change in the **internal electric field** and a **flow of electric current** from one side to the other.

Piezoelectricity is **reversible**: **remove the stress** and the **material returns to its normal state**, while an **externally generated electric field** applied to the material will cause it to **expand** or **contract** slightly.

APPLICATIONS

Timekeeping
Quartz watches generate a **precise time signal** through **piezoelectric oscillations**: an electric field compresses the crystal, which then relaxes and **releases a pulse of current** itself. The **timing** of these oscillations is controlled by the crystal's **fundamental frequency**.

Gas lighters
Lighters generate an **ignition spark** by **striking a piezoelectric crystal with a hammer**. A large **potential difference** is created between the crystal and a nearby **conducting plate**, and **discharged** when a **spark bridges the gap**.

Sonar and ultrasound
Changing the shape of piezoelectric crystals inevitably creates **sound waves**. **High-frequency electric fields** can produce penetrating **ultrasound** used in **sonar** and **medical** applications. The piezoelectric crystals are then **compressed by returning sound waves**, producing **electrical signals** that can be **decoded into images**.

PIEZOELECTRIC SUBSTANCES

Many materials exhibit piezoelectric behaviour. These include:

- A wide range of **natural crystals** such as **quartz**, **topaz**, and **tourmaline**.
- **Natural substances** such as **bone**, **wood**, **silk**, **proteins**, and **DNA**.

SEMICONDUCTORS

Semiconductors are special materials with properties midway between those of electrical conductors and insulators – they can be used to create components in which electricity flows one way and is blocked from the other direction.

SEMICONDUCTOR MATERIALS

Semiconductors are typically made from a **base element** such as **silicon** or **germanium** – both elements with a **half-filled outer shell of electrons**.

Silicon and germanium both form **crystal lattices** in which **charge is unevenly distributed** – **loose electrons cluster** in some places to give them a **negative bias**, while the **holes left by their absence** have a **net positive bias**.

Negative "**n-type**" and positive "**p-type**" areas can be **artificially created** by adding **small amounts of other elements** that help to

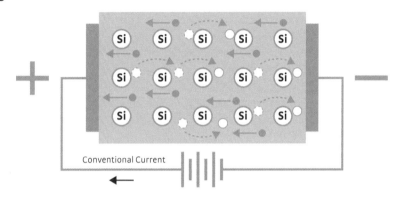

Conventional Current

concentrate or **repel charge** – a process called **doping**.

Both types of bias can be made to **migrate** through the material,

meaning that semiconductors can be treated as having **charge carriers** that **flow in either direction**.

DIODES

The diode is the **simplest semiconductor device** – a component that acts as a **valve**, allowing **electricity to flow freely**

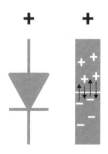

in one direction but **blocking its passage in the other**.

- A p-n junction has an **excess of electrons on one side**, and an **excess of holes on the other**.
- Current **passes freely** from **n-type** to **p-type** material.
- **Concentration of electrons** in n-type material **blocks the acceptance of any more**, so **current cannot return**.

Diodes can be designed with either "**forward bias**" shown here, or "**reverse bias**" (in which the **preferred direction of current flow is reversed**). Their most common use is as a stage in "**rectifying" alternating current** for use in **direct-current circuits**.

They can also be designed so that **current will suddenly start to flow** when a certain **potential difference** between the **two sides of the p-n junction** is reached.

ANALOGUE AND DIGITAL ELECTRONICS

Analogue electronics is concerned with currents that vary smoothly and continuously in strength or direction –
but it's also possible to apply electrical techniques to currents that step abruptly between a limited set of values.

DIGITIZING DATA

Digital electronics break down the behaviour of a **continuously varying electric current** into a **stream of numbers**:

- The **strength** of the **original signal** is sampled at **high frequency**.
- This value is **normalized** in relation to a **specific range of numbers** – for instance, 156 on a range of 0 to 255.
- This number is **converted** into a **stream of "0"s and "1"s** in the **binary** or base-2 numbering system: 156 = 10011100. Each 0 or 1 value is known as a **bit of data**, and a stream of 8 bits makes a **byte**.
- The signal is then **processed** or **transferred** as a **stream of electrical pulses** with **well-separated values** indicating 1 and 0.
- The **original analogue waveform** of the signal may be **reconstructed** at the other end using a **digital to analogue converter (DAC) circuit**.

BINARY NUMBERS

Binary is a **place-value counting system** just like the more familiar **decimal**, except instead of having **ten possible digits** (0 to 9), it has just **two** (0 and 1):

Decimal		Binary			
"10s"	"1s"	"8s"	"4s"	"2s"	"1s"
	0				0
	1				1
	2			1	0
	3			1	1
	4		1	0	0
	5		1	0	1
	6		1	1	0
	7		1	1	1
	8	1	0	0	0
	9	1	0	0	1
1	0	1	0	1	0

ANALOGUE VS DIGITAL

Analogue

Digital

Analogue current can directly reflect the strength of **real-world phenomena** (e.g. **sound waves**).

Noise (random fluctuations) and **interference** can **disrupt** analogue signals so that **original precise values** are **lost**.

In **digital signals**, the **difference between binary 0s and 1s** is large enough to **avoid confusion**, so **noise has no effect**.

Digital 0s and 1s can be **processed** in electronic components such as **transistors** and **logic gates**.

Increasing sample rate and **range** produces a **better copy** of the **original analogue signal** (8 bits = 256 possible values, but 16 bits = 65,536 possible values).

ELECTRONIC COMPONENTS

A wide range of different electronic devices can be used to process digital data, mimicking the performance of analogue electrical components but also achieving far more complex tasks.

TRANSISTORS

A transistor is an electronic device. Current **flows** or is **blocked** between **one pair of these terminals** depending on the **current at a third terminal**, or the **voltage across a third and fourth**.

The transistor has two functions:

- **Switch**: Output is turned **on** or **off** depending on **applied current/voltage**.
- **Amplifier**: Output **mirrors current input** but provides a **current strength of whatever level is desired**.

TRANSISTOR TYPES

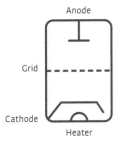

Thermionic triode: The **first transistor** – a **glass vacuum tube** with **anode** and **cathode electrodes** and an **intervening grid** whose state controlled the flow of current.

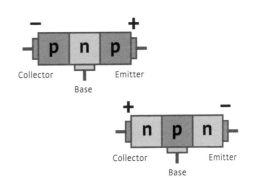

- **Bipolar junction transistor**: A sandwich of **n-type** and **p-type semiconductor materials** arranged **npn** or **pnp**.

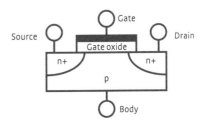

- **Field-effect transistor**: **Flow of current** between **source** and **drain terminals** is controlled by an **electric field (potential difference)** between **gate** and **body**.

LOGIC GATES

Electronic components such as **transistors** and **diodes** can be **combined** to **perform simple logical operations** – making a **comparison** between **input values** to produce an **output** according to a **simple logical test**. Known as logic gates, these units are at the heart of **modern computer technology**:

AND	Produces output current if **two input currents are present**.
OR	Produces output current if **either input receives a current**.
NOT	Produces output only if there is **no current to its single input terminal**.

NAND	Produces output current **unless two input currents are present**.
NOR	Produces output current only if **neither input current is present**.

SUPERCONDUCTIVIITY

In certain extreme conditions, some materials can conduct electricity without resistance – an electrical equivalent of the strange superfluid behaviour seen in some low-temperature materials.

HOW SUPERCONDUCTIVITY WORKS

Superconducting materials act as **perfect electrical conductors** with **zero resistance**, and also **actively expel magnetic fields** that **attempt to pass through them**.

Superconductivity occurs in many materials at **very low temperatures**, within **30°C of absolute zero**.

Several mechanisms seem to be at work, of which the best understood is the **Bardeen-Cooper-Schrieffer (BCS) model**:

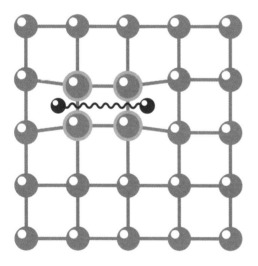

- At **low temperatures, electrons bond weakly together** to form "**Cooper pairs**".
- The paired particles behave as **bosons** (see p.137) – unusual particles that **can all take on identical properties** rather than necessarily existing in a jumble of different states. This **reduces their friction** with each other and the surrounding material.

APPLICATIONS

Superconductor materials can be used in **high-powered electromagnets** capable of generating **powerful magnetic fields**. These are integral to a wide range of **advanced technologies**:

- **Triggering oscillations** required by **Magnetic Resonance Imaging (MRI) medical scanners**.
- **Boosting charged particles** to **high speed** in **particle accelerators**.
- **Creating frictionless bearings** for the **Japanese Shinkansen bullet train**.
- **Generating powerful magnetic fields** for **compressing materials** in **prototype nuclear fusion reactors**.

RAISING THE TEMPERATURE

Finding a **room-temperature superconductor** that **works in normal conditions** at **temperatures above 0°C** is a holy grail of modern physics – such a material would **revolutionize energy transport and production**, as well as **computing** and **other fields of electronics**. However, practical room-temperature superconductors are **still a long way off**:

- **Lanthanum decahydride**: (LaH_{10}): superconducting below -23°C, but only under **intense pressures**.
- **Mercury barium calcium cuprate** ($HgBC$-CO): Superconducting below -135°C at **normal atmospheric pressure**.

INTEGRATED CIRCUITS

Compact modern electronic devices rely on the miniaturization of individual components to the scale of nanometres (billionths of a metre), uniting millions or billions of elements on a single chip of semiconductor material.

CHIP MAKING

Modern integrated circuits (ICs) are mostly built around **field-effect transistors** (see p.107) made using a technique known as CMOS (**complementary metal–oxide–semiconductor**).

Very small amounts of different substances are "**etched**" onto the surface of an **underlying semiconductor substrate** using a **specialized printing technique** known as **photolithography**. **Ultraviolet light** shines through a **semi-opaque mask** and triggers **chemical reactions** on the areas it illuminates.

Different areas of the underlying semiconductor substrate are **doped** with various **metals** to produce **p-type** or **n-type** areas.

Insulating materials and **conducting metals** such as **aluminium** are laid down to **join up** the **semiconductor areas** and create **logic gates**.

CMOS circuits generate **little noise** and have **low power consumption**, which allow them to be **packed onto an IC at very high densities**.

CHARGE-COUPLED DEVICES

CCDs are specialized ICs used in **electronic imaging**:

- Individual **photons of light** striking an **array of pixels** (picture elements) produce **electrons** due to the **photoelectric effect**.
- **Negative charge** builds up in proportion to the **number of photons**, but is **trapped on one side of an underlying capacitor** in "**potential wells**".
- When the CCD is **read** at the end of a **set exposure time**, the **first pixel** in the array **drains its charge to an amplifying circuit**, which **converts it to a voltage**, while the **remaining pixels pass their charges along the chain to their neighbours**.
- **Repeating the process** transforms the **pixel brightnesses** into a **sequence of voltages** that can then be **electronically processed**.

MOORE'S LAW
Coined by **Intel co-founder Gordon Moore** in 1965 (and revised a decade later), Moore's law states that the **density of transistors in integrated circuits doubles**, on average, **every two years** as **technology improves**. It accurately forecast the huge **boom in computing power** as ICs grew more powerful in the late twentieth and early twenty-first century, though it has **tailed off** since the 2010s as the **limits of current manufacturing technology**, with components just a few nanometres across, are approached.

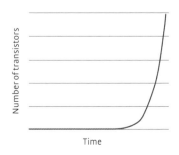

VACUUM TUBES

Vacuum tubes are a varied family of devices that manipulate electric current using a pair of electrodes in an evacuated glass tube, across which a large potential difference is created. They have a wide range of uses.

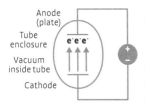

OPERATING PRINCIPLE

In general, a vacuum tube consists of a **negative electrode** (**cathode**) and a **positive electrode** (**anode**) **separated by a near-perfect vacuum**. This allows a **high voltage** to be created between them **without the air between them ionizing** to **conduct electricity in sparks**. Instead, **electrons** are **emitted directly from the cathode** and **flow to the anode**.

THERMIONIC TUBES

- In thermionic tubes, a **stream of electrodes** is generated as the **cathode is heated** and the **additional energy allows them to break free**.
- Thermionic triodes have a **third grid-shaped electrode** between **anode** and **cathode**: its **relative voltage controls the flow of current**.
- **Cathode ray tubes** (CRTs) use **electric** or **magnetic fields** to **direct the stream of electrons** from a cathode called the "**electron gun**" onto a **screen coated with phosphor dots**, creating a glowing dot. **Fast-changing fields** can be used to **draw and redraw images faster than the human eye can register**, as seen in **CRT displays** and **old-fashioned tube televisions**.
- Adding a small amount of a specific **inert gas** (such as **neon**) back into a vacuum can create a **discharge tube** that produces **glowing light**.

PHOTOELECTRIC TUBES

In photoelectric tubes, **electrons** are produced from the **cathode** thanks to the **photoelectric effect** – **photons of light** provide the **energy** to **boost the electrons out of atoms on a negatively charged photocathode**:

- **Photomultiplier tubes focus the electrons into a beam** before **bouncing them down a "corridor" of electrodes** called **dynodes**. Each dynode has a **higher positive voltage than the last**, and the **resulting electric field encourages electrons to become dislodged**, until a **single electron** released at the **photocathode** gives rise to an **avalanche eventually reaching the positive anode**.
- **Vacuum phototubes lack amplification**, but can **generate a current** at the **anode** in **direct proportion** to the **light hitting the photocathode**. Though **mostly obsolete**, they can be **tuned to very specific frequencies** depending on the photocathode material.

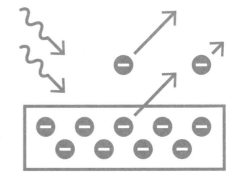

DISCOVERING SUBATOMIC PARTICLES

Until the late nineteenth century, most scientists believed that atoms were the simplest objects in nature. Then a series of discoveries revealed that each atom was itself made of subatomic particles.

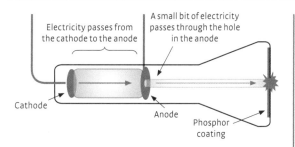

Electricity passes from the cathode to the anode

A small bit of electricity passes through the hole in the anode

Cathode

Anode

Phosphor coating

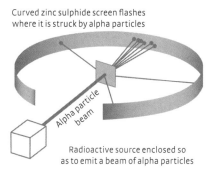

Curved zinc sulphide screen flashes where it is struck by alpha particles

Alpha particle beam

Radioactive source enclosed so as to emit a beam of alpha particles

CATHODE RAYS

In 1897, **J. J. Thomson** carried out investigations into the mysterious cathode rays (rays of an **unknown kind** produced at the **negative electrode** in **glass vacuum tubes** with a **high voltage difference** between its ends).

- The rays **heated** other objects they **struck**, suggesting that they were **particles** transferring **kinetic energy**.
- By considering the **distance they could travel** before being **stopped** by **air**, Thomson concluded that they were much **smaller than atoms**.
- By measuring their **deflection** in **electric and magnetic fields** he showed that they carried a **negative electric charge** and even estimated their **charge-to-mass ratio**.
- The rays exhibited the **same properties** regardless of the **elements** used in the **cathode** producing them.

DISCOVERING THE NUCLEUS

In 1908, **Ernest Rutherford**, **Hans Geiger**, and **Ernest Marsden** conducted the famous **"gold foil"** **experiment**, using **radioactive alpha particles** (see p.120) to probe the atom's **internal structure**.

If the **plum pudding model** was **correct**, the sheet should have presented a **uniform barrier** to the alpha particles. Instead:

- Most particles passed through in a **straight line**.
- Some were deflected at **wide angles**.
- A few were **bounced back towards the source**.

The only explanation was that the atom's **mass** and **positive charge** are **concentrated** in a **nucleus** at its **centre**.

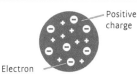

Positive charge

Electron

Positively charged nucleus

Negatively charged electrons

Thomson concluded that the **electrons**, as they became named, were **tiny, low-mass particles** found inside **every atom**.

THE PLUM PUDDING MODEL

The **earliest models** of the atom imagined **electrons** scattered at **random** through the **positively charged body** of the atom, like plums in a pudding.

THE PLANETARY MODEL

Based on his experiment, Rutherford devised a **planetary model** of the atom: **electrons** follow different **orbits** around a **central nucleus** like **planets** around the **Sun**, with most of the atom being just **empty space**.

THE BOHR MODEL

In 1913, Niels Bohr published three papers that established a powerful model for understanding atoms. Although we now know the reality of atoms to be more complex, Bohr's model remains widely used even today.

THE RYDBERG FORMULA

Bohr's model was based on a **law** that had been established in **spectroscopy** since the 1880s. Generalizing from the widely studied **Balmer series**, the **pattern of spectral lines emitted by hydrogen** could be described by the **Rydberg formula**:

$$\frac{1}{\lambda_{\text{vac}}} = R_H \left(\frac{1}{n_1^2} - \frac{1}{n_2^2} \right)$$

- l_{vac} is the wavelength of emitted light.
- R_H is a constant, the Rydberg constant for hydrogen.
- n_1 and n_2 are two series of small whole numbers (1, 2, 3 ...).

ENERGY LEVELS

Bohr suggested that the **Rydberg series** arose because **electrons** are **limited to particular values of angular momentum**, and therefore **particular distances from the nucleus**:

- An electron has an **energy** associated with its **movement** and its **distance from the nucleus**.
- **Orbits closer to the nucleus** have **lower energies than those further out**, so electrons will tend to **"fall"** towards the **closer orbits** if they can.
- However, each orbit has a **maximum capacity for electrons**.
- Electrons can **jump** into **higher orbits** if an atom is supplied with **external energy** (such as **heat** or **light**), but will rapidly fall back to the **lowest available energy level**.

Hence **atoms** will **absorb** and **emit light** at **specific energies** and **wavelengths.**

TRANSITIONS

The **light emitted** or **absorbed** by an **electron moving between two energy levels** in **Bohr's model** obeys the simple equation:

$$\Delta E = h\nu$$

- ΔE is the **energy required** or **released** in the transition.
- ν is the **frequency of emitted light** (or other **electromagnetic radiation**).
- h is a constant (**Planck's constant**).

AN EARLY SUCCESS

Bohr's model proved its worth when he **successfully used it** to explain a group of **dark "absorption lines"** seen in the **light of very hot stars**. He showed that this "**Pickering series**" was created in a star's **atmosphere** when the **single remaining electron in ionized helium** (He⁺) absorbed **various frequencies** of **high-energy radiation** escaping from the star's **luminous surface.**

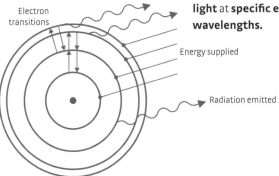

Electron transitions

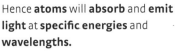

Energy supplied

Radiation emitted

NUCLEONS

Any atomic nucleus is made of two types of subatomic particle – protons and neutrons.
The balance between these two particles determines the atom's nuclear properties.

PROTONS

A **single proton** has a **mass** of approximately

$$1.67 \times 10^{-27} \text{ kg}$$

or **1 atomic mass unit** (amu). This is about the same as **1,836 electrons**.

A proton's **diameter** is

$$1.7 \times 10^{-15} \text{ m}$$

(1.7 femtometres or millionths of a billionth of a metre).

A proton carries an **electric charge** of

$$+1.602 \times 10^{-19}$$

coulombs, a positive "**elementary charge**" equal and opposite to the **charge on the electron**.

On a more fundamental level, protons are composed of **particles** called **quarks**. Each proton contains **two up quarks** and a **down quark**.

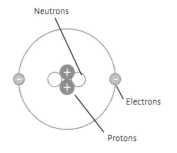

Neutrons
Electrons
Protons

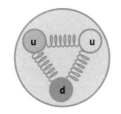

DISCOVERING THE PROTON

In 1917, **Ernest Rutherford** discovered that **atoms** of many different **elements** contained a **positively charged particle** with properties **identical to that of a hydrogen nucleus**. He named this building block the **proton**.

NEUTRONS

Neutrons are present in the **nuclei of all atoms** except the **simplest form of hydrogen**. They are particles with the **same mass as protons**, but **no electrical charge**, which makes them **harder to detect**.

Atoms usually contain roughly the **same number of neutrons** as they do **protons**, though this can vary.

Neutrons, like **protons**, are composed of **quarks**. A neutron contains **two down quarks** and an **up quark**.

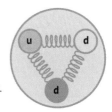

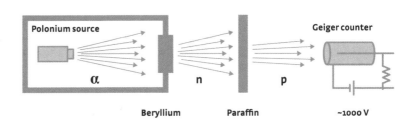

Polonium source · α · Beryllium · n · Paraffin · p · Geiger counter · ~1000 V

DISCOVERING NEUTRONS

In 1932, **James Chadwick** showed that when **radioactive alpha particles** bombarded certain **lightweight metals** such as **beryllium**, they produced a **stream of unknown particles** with a **mass similar to the proton** but **no electric charge**. The particles themselves were **undetectable**, but when they hit a **screen of hydrogen-rich paraffin wax**, high-energy protons were kicked out. The **new particle** was named the **neutron**.

ISOTOPES

Isotopes are atoms of the same element with different masses. The number of protons in the nucleus remains the same (defining their atomic number and fundamental chemistry) but the number of neutrons varies, sometimes affecting their physical behaviour.

EXAMPLE ISOTOPES

Isotopes are usually identified with the **element name** followed by its **atomic mass**, or the symbol *preceded* by the **mass** as a **superscript**. For example, the radioactive isotope **Uranium-238** is also written as ^{238}U. Some isotopes of the **lightest elements**:

HYDROGEN ISOTOPES

$^{1}_{1}H$ Hydrogen $^{2}_{1}H$ Deuterium $^{3}_{1}H$ Tritium

Hydrogen-1 (^{1}H): "normal" hydrogen

Hydrogen-2 (^{2}H): deuterium or "heavy hydrogen"

Hydrogen-3 (^{3}H): tritium

Helium-3 (^{3}He): a light isotope of helium

Helium-4 (^{4}He): "normal" helium

ISOTOPE SCIENCE

Isotopes have found uses in many different branches of science. The **proportions** of different isotopes present are usually calculated through **mass spectrometry** (see p. 57).

In any mixture, **heavier isotopes** will tend to **fall** and **lighter ones rise**. This simple fact means that heavier isotopes will tend to **accumulate in certain parts of the environment** rather than others. During **cold climate conditions**, **water evaporating** from the **ocean** tends to leave behind **rare molecules** containing the **"heavy" isotope oxygen-18**. As some of this water ultimately ends up **layered**, year by year, in the **polar ice sheets**, the **isotopic balance** can reveal **changing climate conditions** stretching back for many thousands of years.

The preference of **plants** for **lighter carbon-12** over the **heavier isotope carbon-13** means that **plant-derived fossil fuels** such as **coal** have a **lower proportion** of ^{13}C than the **environment in general**. A shift in the **atmospheric balance** towards more **carbon-12** over the past century is strong evidence that the **rising amounts of carbon dioxide in Earth's atmosphere** come largely from fossil fuels.

MAKING ISOTOPES

Isotopes are **produced naturally** in three main ways:
- **Nuclear fusion** during the **lives of stars** and in **supernova explosions** (**fusing smaller nuclei** together to make **larger ones**, some of which may be **unstable**).
- **Radioactive decay** of other isotopes – the **fragmentation** of **unstable atomic nuclei** to produce **smaller ones**.
- **Cosmic-ray bombardment** – high-energy particles emitted by the **Sun** and other **astronomical sources** striking and transforming **atomic nuclei**.

In addition, **artificial isotopes** can be made by:
- **Stimulated decay** – radioactive decay events triggered by **bombardment of nuclei** with other particles.
- **Irradiating certain nuclei** in a **torrent of neutrons** within a **nuclear reactor**, leading to the **absorption of neutrons**.
- **Bombarding elements with high-energy particles** produced in **particle accelerators** (used for manufacturing **entirely synthetic heavy elements**).

FLUORESCENCE AND PHOSPHORESCENCE

While most natural materials are visible only through reflected light, a few elements emit light of their own. Fluorescence and phosphorescence are two related effects that are linked to subatomic behaviour within certain atoms.

FLUORESCENCE

Fluorescent materials appear to **glow** with **specific colours** when exposed to **high-energy visible or ultraviolet light**.

- Energetic **blue** or **ultraviolet light** boosts or excites **electrons** in the material's **outer shells** into **higher energy levels**.
- The electrons then **cascade down** through **empty orbits**, emitting **small bursts of energy** at each step until they **return to their original state**.
- The **smaller bursts of energy** correspond to **light of lower frequency** and **longer wavelength** than the light **initially responsible** for the excitation.

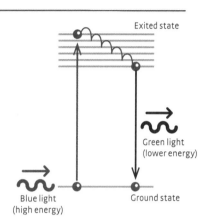

Exited state

Green light (lower energy)

Blue light (high energy)

Ground state

PHOSPHORESCENCE

Phosphorescent materials **glow too weakly to be seen when other light is present** – their secret only becomes clear as they continue to **shine faintly** after **all sources of light have been removed**. After being **excited by light**, **electrons** in these materials take **much longer to return** to their **normal state**.

The delay arises because **returning to the lower-energy state** requires **specific energy jumps** known as **"forbidden" transitions**. Though not actually forbidden, these are very **unlikely to happen** according to the laws of **quantum physics** (see p. 127), and therefore the **emission of light is spread out** over minutes or even hours.

FLUORESCENT LIGHTS

Fluorescent lighting involves passing **electric current** through a **tube** that has had the **air removed** and a **small amount of mercury vapour** introduced to **replace** it. The current excites the **mercury atoms**, causing them to **emit ultraviolet light**. Ultraviolet rays create **visible fluorescence** in a **coating** on the **inside of the lamp**, known as a **phosphor**.

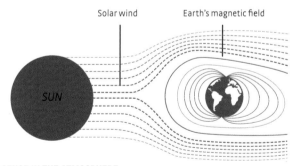

Solar wind

Earth's magnetic field

SUN

FLUORESCENCE IN THE ATMOSPHERE

The beautiful phenomena known as the **aurorae** or **northern and southern lights** are a type of **natural fluorescence** created where **atmospheric gases** are **excited** not by **radiation**, but by **high-energy particles from space**. Electrons trapped by Earth's **magnetic field** are **funnelled down close to the poles**, where they **collide with gases** in the **upper atmosphere**, briefly exciting their electrons and causing them to **emit characteristic glows** as they **return to their normal state**.

LASERS

Short for Light Amplification by Stimulated Emission of Radiation, the laser beam harnesses the natural properties of atoms to produce intense beams of light that find a huge variety of uses.

STIMULATED EMISSION

When an electron is **excited into a higher orbit** by an **injection of energy** from outside, it **rapidly returns** to its previous **lower-energy state** and emits a **burst or photon of light** with a **specific colour**. **Stimulated emission** involves **forcing this change**, rather than allowing it to occur in its **own time**.

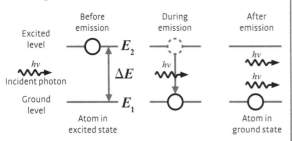

$$E_2 - E_1 = \Delta E = hv$$

- Striking the atom with a **photon of the same wavelength as the transition** pushes the excited electron back down to its **original state**.
- A **second photon** is emitted while the **original survives**.
- The two photons are **identical** not just in their **energy (colour)** *but also* in the **movement of their waves** – they are said to be **coherent**.
- **Beams of coherent light** deliver much more **concentrated** and **precise energy** than **normal light**.

LASERS IN PRACTICE

A laser is a device designed to trigger a **chain reaction** of **stimulated emission**: one **photon** creates a **coherent pair**, which go on to trigger **further emissions** producing **four coherent photons**, and so on. In practice this involves a specific setup:

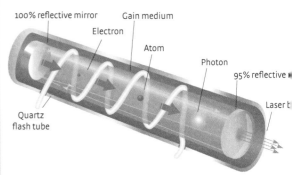

- **Lasing medium**: This is the material whose atoms are excited and that create the emission. **Different lasing materials** create **different wavelengths of light**.
- **External pumping**: Atoms in the lasing medium are excited or "**pumped**" by **intense illumination** or a **powerful electric field**.
- **Optical resonator**: **Mirrors** at either end **reflect photons back and forth** through the lasing medium, forcing the **chain reaction** to build up. One mirror is only **semi-silvered**, allowing some of the **coherent beam to escape**.

LASER APPLICATIONS

Lasers have a huge range of **applications**. As **beams of concentrated light energy**, they can **burn**, **cut**, and **heat materials**, but other uses include:

- **Delivery of energy** at **precisely tuned levels** needed to trigger **atomic transitions**.
- **Precision measurement**.
- **Holography** (see p.80).
- Sending **signals** in **fibre-optic communication**.
- Providing a **precision light source** for use in, for example, **optical data storage** (CDs, etc.) and **scanning equipment** such as **barcode readers**.

ATOMIC TIMEKEEPING

When an electron is temporarily boosted from a low-energy orbit into a higher one, it rapidly falls back to its normal state. The speed at which this happens is so precise that it is used to build the most accurate clocks in the Universe.

BUILDING AN ATOMIC CLOCK

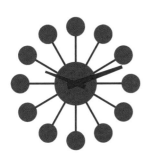

- **Atomic clocks** use **atomic transitions** to create a **high-frequency oscillating electric current** whose **vibrations can be counted**.
- **Vaporised atoms** are **injected into an empty chamber** called a **microwave cavity**.
- **Laser beams** are **fired into the chamber**. Their **photons** are **precisely tuned** to deliver the **energy needed** for the **desired electron transition**.
- Atoms are **excited**, **fall back** to their **ground state** and are immediately **boosted again**, creating a **small but fast-changing electromagnetic field**.
- The cavity **confines** and **amplifies** the **electromagnetic fields**, forcing the **oscillations into sync with each other** and producing **resonant waves** similar to **sound waves**.
- The **resonant waves** are used to drive a **fast-changing current** in the "**timing circuit**".

GLOBAL POSITIONING

The **satellite navigation systems** on which the modern world relies are **critically dependent** on the **accuracy of atomic clocks**:

- A **constellation of satellites** orbits around the Earth, with **several** above the horizon for **any observer at any time**.
- **Atomic clocks** on the satellites **regulate time signals** they broadcast.
- A **receiver unit** detects the **signals** from **multiple satellites** and **compares them** with its own **high-precision clock**.
- The **delay to the time signal from each satellite** is an **indicator** of its **relative distance** from the **observer**.
- The receiver uses a **built-in ephemeris** (a model of the various **satellite orbits**) to **calculate** its **precise location** based on the **distance at that moment to the various satellites**.

TIMEKEEPING ELEMENTS
Common **elements** used in atomic clocks include **hydrogen**, **caesium**, **rubidium**, and **strontium**.

TIME REDEFINED
Since 1967, the **second** has been **officially defined** as the **duration of 9,192,631,770 cycles** of **excitation** and **emission** for a particular **electron transition** in atoms of **caesium-133**.

CHERENKOV RADIATION

Everybody knows that nothing can travel faster than light – but that's not strictly true. While nothing can travel faster than light in a vacuum, it's possible to break the light barrier when light itself is slowed down. When that happens, one result is the effect known as Cherenkov radiation.

HOW IT WORKS

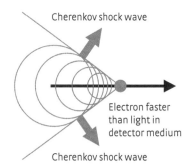

Cherenkov shock wave

Electron faster than light in detector medium

Cherenkov shock wave

Cherenkov radiation is the **light equivalent of the sonic boom** heard when a plane goes through the **sound barrier**. It **only occurs** when the particle is carrying an **electric charge**, and the **medium through which it passes** (called the **dielectric**) has a structure that can be affected by **electric fields**.

- The **electric field** around the particle **polarizes the dielectric** around it for a **brief moment** before it **moves on**.
- As the **dielectric** returns to its **normal state**, it **loses energy** and **emits a burst of light**.
- Because of the particle's **high speed**, the **light-emitting region** forms a **cone-like shock front**, along which **light emissions** are **coherent**. Instead of **interfering** with each other and **disappearing**, the **light waves reinforce** and **become visible**, usually as a **bluish glow**.

DISCOVERY

Pavel Cherenkov discovered the **radiation** that bears his name in 1934 while carrying out **experiments** that **bombarded a water flask with radioactive particles**. Some of the **beta particles (electrons)** were able to **penetrate the walls of the flask** and **pass through water** at **speeds greater than 225,000 km/s** (the speed of light in water), producing a **tell-tale blue glow**.

NUCLEAR GLOW

Cherenkov radiation creates a characteristic blue glow around the **cores of nuclear reactors**, where it is caused by **high-speed particles escaping into the surrounding water**.

COSMIC RAY OBSERVATORIES

Some of the world's strangest **observatories** consist of **huge grids of Cherenkov detector tanks** spread across hundreds of square kilometres. These arrays are designed to **detect** the "**air showers**" of **particles** produced when **mysterious high-energy cosmic rays** from **deep space** enter the **atmosphere** and **collide with air molecules**. The resulting **shower of particles** can spread across a wide area, and the **time delay** between **particles being detected** in **different parts of the array** allows astronomers to calculate the **trajectory of the original particle**.

DETECTING EXOTIC PARTICLES

Astronomers and **particle physicists** use **Cherenkov radiation** to track down some of the most **elusive particles in the Universe**.

- **Sealed tanks** of water or other liquid **lower the speed of light**.
- **High-speed particles** pass through the liquid unaffected at close to *c* and **leave a trail of Cherenkov radiation**.
- **Sensors** around the edges of the chamber are triggered by the **passage of the light**.
- For the **deepest-penetrating particles** such as **neutrinos** (see p.145), **detector chambers** can be **shielded** through **burial underground** (for instance in repurposed mines)

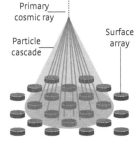

Primary cosmic ray

Particle cascade

Surface array

118

NUCLEAR PHYSICS

The first clues to the complexity of the atomic nucleus came from the discovery of strange new types of radiation in the late nineteenth century. Within a generation, these opened up a new world of nuclear science with unimaginable consequences.

DISCOVERING RADIOACTIVITY

In 1896, reports of **Wilhelm Röntgen's recent discovery** of **highly penetrating X-rays** inspired **Henri Becquerel** to investigate whether **phosphorescent materials**, such as certain **salts of the heavy element uranium**, produced **similar rays**.

Becquerel found that the salts **did indeed emit rays** that **passed through solid objects** and **fogged photographic film**. Further experiments showed that the **phosphorescent salts did not depend on an external source of energy**. They continued to emit rays even if kept in the dark for long periods. Furthermore, **non-phosphorescent uranium compounds also emitted rays**.

Marie Curie discovered that **uranium salts ionized the air around them**, and that the **amount of activity** seemed to depend **solely on the amount of the compound present**. With her husband **Pierre**, she went on to discover new "**radioactive**" elements far more active than **uranium**.

In 1899, **Ernest Rutherford** studied the **penetrating power** of the new emissions and how they could be influenced by **electric and magnetic fields**. This led him to identify **three distinct types of radiation**, now known as **alpha**, **beta**, and **gamma**.

PUSHING THE BOUNDARIES

Alpha, beta, and gamma radiations **occur naturally**, but in the early twentieth century physicists learned to **trigger other changes to the nucleus** through **artificial means**:

- **Neutron emission:** Bombardment of **lightweight metals** with **other radioactive particles** forces them to **release neutrons**.
- **Nuclear fission:** In the right circumstances, **heavy unstable atomic nuclei** can be made to **split**, producing two large "**daughter isotope**" nuclei and a large amount of **energy** instead of simply **decaying through natural radiation**. In 1938, **Otto Hahn** and **Lise Meitner** successfully **split uranium atoms** to produce **krypton** and **barium**.
- **Artificial radioactivity:** It's also possible to turn **previously stable isotopes** into **radioactive ones** by **bombarding them** with **alpha particles**. In 1934, **Irene and Pierre Joliot-Curie** discovered that **neutrons** absorbed by the **stable nuclei** of elements such as **boron** and **aluminium** rendered them prone to **instability** and **decay**.

TYPES OF RADIOACTIVITY

Radioactive atoms produce three distinct types of radioactivity, known as alpha, beta, and gamma radiation (from the first three letters of the Greek alphabet). The three different emissions have little in common, except for their link to radioisotopes.

ALPHA, BETA, AND GAMMA

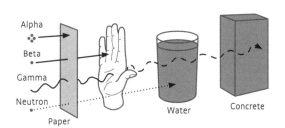

Alpha · Beta · Gamma · Neutron · Paper · Water · Concrete

The three different forms of **radioactive emission** differ in their **penetrating power**, **effects on other materials**, and even their **essential nature**.

- **Alpha** (α) particles are **identical** to the **nucleus** of a **standard helium atom**. Containing **two protons** and **two neutrons**, they are relatively **heavy, slow-moving**, and **easily stopped** by even a **paper-thin barrier**.
- **Beta** (β) particles are **lighter, faster-moving**, and **more penetrating** than alpha radiation. They can carry a **negative** or **positive electric charge**. The **negatively charged form** (β⁻) is simply the familiar **subatomic electron particle**, while the **positively charged version** (β⁺) is a **positron**, the **electron's antimatter equivalent**.
- **Gamma** (γ) rays, as their name suggests, are **not particles** but a **form of electromagnetic radiation** with **exceptionally high energy** (see p. 69). They **travel at the speed of light** and have the **greatest penetrating power of all**, requiring **heavy steel**, **lead**, or **concrete shielding** to **block** them.

ALPHA APPLICATIONS

Since their discovery, **alpha particles** have found a **host of unexpected uses**. Their **limited range** and the **ease with which they are blocked** allows them to be used in **close proximity to humans**:

Smoke detectors use alpha emission from the **americium-241 isotope** to **ionize air** and pass a **current** that is **interrupted by smoke particles**.

Radioisotope Thermoelectric Generators (RTGs) make use of the **heat** released in **alpha decay** to **generate electricity**. They are used to **power spacecraft** and **satellites** where **solar panels** are not possible, and in the past were even used to power **heart pacemakers**.

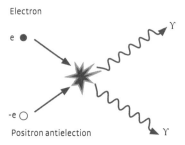

Electron · e · -e · Positron antielection

NATURAL ANTIMATTER SOURCE

Whenever **antimatter particles** come into contact with **normal matter**, they usually **annihilate** in a **burst of energy** that is released as **gamma rays**. **Positrons** emitted by **radioisotopes** such as **carbon-11**, **nitrogen-13**, and **potassium-40** are usually **destroyed** in this way **moments after their creation**, but **shielded sources** kept in **isolation** can be used to **provide antimatter** for **laboratories** and **medical applications**.

RADIOACTIVE DECAY (SERIES/CURVES)

Radioactive emission occurs until an unstable atomic nucleus achieves a state of long-term stability. The three different forms of emission all represent different means for the nucleus to reach a more stable state at a particular moment.

DECAY SERIES

Unstable atomic nuclei tend to have an **imbalance of protons and neutrons** – which in practice usually means **more neutrons than protons**. To reach a **more balanced state**, the **nucleus** can use three possible **decay mechanisms:**

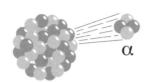

- **Alpha emission:** The nucleus **sheds two protons** and **two neutrons** through a **"quantum tunnelling"** process in which an **energetic cluster** of **nucleons breaks free of the forces**

binding the overall nucleus together.

- **Beta-minus (β⁻) emission:** A **neutron** in the **nucleus** spontaneously **transforms** into a **positively charged proton**, with a **negative electron** produced to **balance the overall charge**.
- **Beta-plus (β⁺) emission:** In some rare situations where

an **isotope is imbalanced in favour of protons**, a **proton** may spontaneously **transform** into a **neutron**. In this case the **overall electric charge** is **balanced** by the **emission of a positron** (a **positively charged "antimatter electron"**).

After any of these changes, the **nucleus rearranges itself** into its **lowest possible energy state**, leading to the **shedding of excess energy** through **gamma radiation**.

DECAY CURVES

Any particular **radioactive decay event** is inherently **unpredictable** – a particular atom **may or may not decay** in a **given timeframe**. With large samples, however, **statistical rules** can be applied:

- A **certain proportion** of **atoms** in a sample **will decay** in a given time.
- After a certain amount of time, the **amount of the original radioisotope will have halved**.
- This is known as the **half-life** for a particular radioisotope.
- After a **further half-life**, a **quarter of the original radioisotope** will remain.

However, the **decay of one radioisotope** may lead to its **(temporary) replacement** by others from **further down the decay series**.

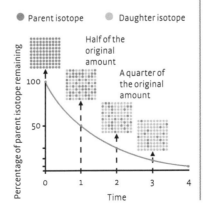

EXAMPLE DECAY CHAIN OF URANIUM-238

Isotope	Decay type	Half-life
→ Uranium-238	a	4.47 billion years
→ Thorium-234	β⁻	24.1 days
→ Protactinium-234	β⁻	1.16 minutes
→ Uranium-234	a	245,000 years
→ Thorium-230	a	75,400 years
→ Radium-226	a	1,600 years
→ Radon-222	a	3.82 days
→ Polonium-218	a	3.10 minutes
→ Lead-214	β⁻	26.8 minutes
→ Bismuth-214	β⁻	19.9 minutes
→ Polonium-214	a	164.3 microseconds
→ Lead-210	β⁻	22.2 years
→ Bismuth-210	β⁻	5.0 days
→ Polonium-210	a	138.4 days
→ Lead-206	−	stable

NB: While this is the **principal decay chain**, small amounts of the **parent isotope** may take **different decay pathways** at **several points**.

RADIOMETRIC DATING

Increased understanding of radioactive decay has opened the way for ingenious methods of dating rocks and even living materials. The most famous of these is radiocarbon dating, widely used in archaeology.

RADIOCARBON DATING

All living things incorporate **carbon** from their environment, which includes a small amount of **carbon-14**, a **mildly radioactive isotope** with a **half-life** of 5,730 years, present across Earth's **atmosphere**, **oceans**, and **land surface**.

- Carbon-14 in the atmosphere is steadily created when **cosmic rays strike molecules of carbon dioxide** – hence the **ratio** of ^{14}C to ^{12}C is kept **more or less constant**.
- Living things **constantly exchange carbon** with their environment, ensuring they contain a **similar proportion of carbon-14**.
- When an organism **dies**, the exchange of carbon-14 **stops**. **Radioactive decay** gradually **diminishes** the **proportion** of carbon-14 in **surviving material**, even as the remains decay.

DATING THE EARTH

While **carbon dating** is useful for **recent organic remains**, other **radioactive decay series** are useful for **dating older natural materials**. The most important of these methods is **uranium-lead dating**. This relies on the **uranium decay series** (see p.121), in which the **half-life of uranium-238** itself is 4.47 billion years:

- **Geologists** look for **zircon crystals** that incorporate **uranium** when they form but whose **chemistry** rejects **lead**. These crystals can **resist natural destruction** even over **billions of years**.
- **Any lead found** in the crystal must have come from the **decay of uranium**, so the current ratio of **lead to uranium** indicates what **proportion** of the sample has **decayed**.
- Since the **half-lives** of ^{238}U and other **isotopes in the decay chain** are **precisely known**, it's possible to **estimate the age of rocks** to **less than a 1 per cent margin of error**, even if they are billions of years old.

CARBON-DATING WRINKLES

- Each successive **half-life** diminishes an **already-small proportion of carbon-14**, so **measurements** become **statistically less accurate for older materials**.
- In practice, this means that carbon dating is **only useful for organic materials up to around 60,000 years old**, and its **precision falls off rapidly in older samples**.
- The proportion of environmental carbon-14 is **affected by Earth's climate**, the **intensity of cosmic rays** and other phenomena – including **modern technology**.
- Measurements must therefore be **calibrated against a historical record**, drawn from **ancient atmospheric samples trapped in ice cores**.

EARTH'S INTERNAL ENGINE

Although Earth retains some of the **heat generated** when the **rocks that formed it** came together in **violent collisions** more than 4.5 billion years ago, much of its **geological activity** has been **powered** by the **slow**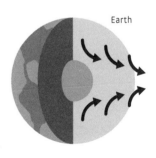

Earth

decay of uranium and other **radioactive isotopes** deep within the planet. These isotopes were themselves created by nuclear fusion in exploding stars before our solar system was born.

NUCLEAR ENERGY

The energy released by radioactive decay is linked to the "binding energy" that holds the atomic nucleus together. This opens two routes to energy from nuclear processes: either by splitting nuclei apart, or by forcing them together.

NUCLEAR BINDING ENERGY

Binding together several particles to make an **atomic nucleus** is just like binding together **atoms** or **molecules** when a **liquid freezes** or a **vapour condenses** – because **individual particles require less energy** than they did in their **previous state**, they can **shed the excess**. The **binding energy** of a **particular nucleus** is the energy that would (theoretically) have to be supplied in order to **split it apart** into its **individual protons** and **neutrons**.

- Because **energy** and **mass** are **equivalent** according to **Einstein's famous equation** $E = mc^2$, the **binding energy "given up" by** a particular nucleus manifests itself as a **difference in mass**.
- The **shortfall** between the **mass of an atomic nucleus** and the **mass of the same number of individual nucleons** (protons and neutrons) is known as the **mass defect**.
- For example, the **combined mass of six individual protons** and **six neutrons** is approximately **0.8 per cent larger** than the mass of a combined carbon-12 nucleus.

FISSION OR FUSION?

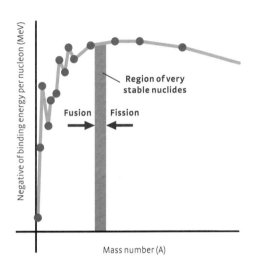

Mass number (A)

The **binding energies** of atomic nuclei **vary more than would be expected** from simply calculating their **individual number of nucleons**, and can be represented on a graph of the **binding energy per nucleon**.

For **elements up to iron** (atomic number 26), binding energy **increases**, as you might expect. Up until this point, **fusing lighter nuclei** together to make **heavier ones** will **always result** in a **release of energy**.

For elements heavier **than iron**, however, the binding energy per nucleon **decreases**. This is because, as the **nucleus grows larger**, **repulsion** between **protons at its fringes** grows **stronger** in comparison to the very **short-range strong nuclear force** that **attracts the nucleons** together. As a general rule, this means that **fusing heavy elements** together **absorbs energy** and **cannot be a viable source of energy**. **Fission** or **splitting apart** of **heavy elements** to form **lighter ones**, however, **does release energy**.

FUSION ENERGY

Nuclear fusion is the power source of the stars. It releases energy not by splitting apart the nuclei of heavy elements, but by joining together light ones. It also has the potential to be a source of limitless clean energy on Earth.

FUSION IN THE STARS

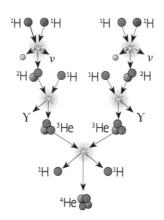

ν Gamma ray
Y Neutrino
● Proton
● Neutron
○ Positron

Most **stars** are powered by the **simplest form of nuclear fusion** – the **combination of individual hydrogen nuclei** (protons) to form **nuclei of helium**:

- Conditions in the **heart of stars** like the **Sun** reach **temperatures** of 15 million °C and **pressures** a quarter of a million times that of **Earth's atmosphere**. This **strips away electrons** and allows **positively charged atomic nuclei** to **collide** at **speeds high enough to overcome their mutual repulsion**.
- **Hydrogen fusion** can follow one of **two processes**: the **simple proton-proton (PP) chain** in relatively **small stars** like the **Sun**, and the **CNO cycle** in **heavier stars** with even **hotter cores**. Both ultimately result in the formation of **helium**.
- While the **amount of fusion** released by the **manufacture of an individual helium atom** is the **same in each case**, the **CNO cycle** (in which helium atoms are "**built up**" inside **heavier nuclei** before being released) **takes place much more quickly than the PP chain process**.
- Hence **heavyweight stars shine much more brightly** and "**burn through**" the **hydrogen in their cores** in **millions** of years, rather than **billions** for more sedate stars like our **Sun**.
- Once **useful hydrogen** in the **core** of a **Sun-like star** has been **exhausted**, the star goes through a **number of changes** that **increase its core temperature and pressure** still further. This allows it to **fuse helium nuclei** together (and may permit **further stages of fusion** before the star is exhausted).

FUSION ON EARTH

Attempts to create **nuclear fusion on Earth** depend on **reproducing conditions close to those in the Sun**:

- Fusion reactor is a **doughnut-shaped torus vessel**.
- Powerful **electromagnets compress** and **confine charged ions** inside the vessel without touching the walls.
- **Lasers**, **magnets**, or **electric fields heat the fusion material** to **millions of degrees**.
- Reaching the **extreme conditions** needed to **fuse individual protons** together is impractical – instead reactors aim to replicate the more **achievable fusion** of **deuterium** and **tritium**, the **"heavy" hydrogen isotopes** that appear **further down the PP chain**.

FISSION POWER

Nuclear reactors harness nuclear chain reactions in a controlled manner to multiply the energy released by artificially splitting atomic nuclei.

CHAIN REACTIONS

Nuclear power plants rely on a process called **stimulated decay**. While the **precise moment** at which a **single radioisotope** will **naturally decay** is **unpredictable**, it's possible to **force the process** by **striking an unstable nucleus** with another **subatomic particle** (often a **neutron**). Rather than **normal decay**, however, the result is a **fission reaction** that produces **two lighter atomic nuclei** known as "**daughter isotopes**", sometimes accompanied by the **release of other stray particles** such as neutrons.

In a **chain reaction** each **fission event** releases **neutrons** in addition to the **daughter isotopes**. These **strike other nearby atoms of the original parent isotope** and **trigger their decay in turn**.

If a **fission event** releases **more than a single neutron**, then a **runaway cascade** can be triggered, releasing **potentially dangerous** amounts of **heat**, **energy**, and **radiation**. **Moderators** are materials (such as **graphite**, or **heavy water** made with the **deuterium isotope**) that are used to **surround the fissile material**. They help to **absorb** or **slow down** neutrons, **reducing the rate** at which the chain reaction takes place.

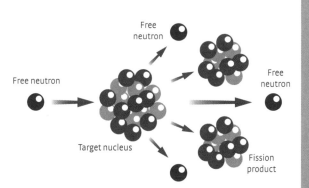

Free neutron

Free neutron

Target nucleus

Free neutron

Free neutron

Fission product

POWER STATIONS

Nuclear power stations use the **energy released by fission** to **heat water** pumped around a **cooling system**. As the water turns to **steam**, it **drives the turbines** that **produce actual electricity**. Since the **first nuclear power plant** became operational at **Oak Ridge, Tennessee** in 1948, several **different designs** have been built, including:

- **Pressurized water reactor**: The vast majority of modern nuclear power plants use **normal water** that is **pumped into the reactor core under pressure** to act as both a **neutron moderator** and a **working coolant** for **generating steam**.
- **Breeder reactor**: This **renewable fuel design** can **generate more fuel than it uses**. An **excess of neutrons** produced by the **fissile fuel** is used to **bombard "fertile" material** (including **waste from other nuclear plants**) and **transform** it into **usable radioisotopes**.
- **Molten salt reactor**: This compact design uses a **liquid salt** as its **coolant**, and **thorium**, rather than **uranium**, as **fuel**. Such reactors avoid many of the **hazards** associated with **traditional nuclear power plants**.

NUCLEAR WEAPONS

In contrast to power stations, nuclear weapons release vast amounts of energy generated by fission or fusion in an instant, converting mass into energy with devastating effect.

ATOMIC BOMBS

The **simplest nuclear weapons** derive their **energy** directly from a **runaway fission reaction**. In practice this means forming a "**critical mass**" of **fissile material** so that a **single decay event** triggers the **start of a chain reaction**.

- **Only certain radioisotopes** (most commonly **uranium-235** and **plutonium-239**) can be **packed densely enough** to power runaway fission – hence **raw materials** must be **processed** or "**enriched**" to **concentrate** them.
- **Atom bombs** use one of two methods to achieve **critical mass**. The simplest option is to use a **gun-like mechanism** to **combine two sub-critical masses into one**. A more sophisticated approach is to **load the weapon with fuel at sub-critical density**, which is then **compressed** by the **firing of conventional explosive "lenses"**.
- A major challenge in **fission weapons** is **preventing them from blowing apart before the fission reaction has run its course**. A "**tamper**" of **dense material** surrounding the **core** can **hold it together** for longer and **reflect neutrons** that would otherwise **escape**.

THERMONUCLEAR BOMBS

Weapons that make use of **fusion** can be **several orders of magnitude more powerful than those that rely on fission alone**. Their **operating principle** is to use the **heat** and **pressure** from a **relatively small fission explosion** to trigger **fusion in a surrounding layer of fusion fuel**.

- **Fusion material** is a **mix** of **deuterium** and **tritium hydrogen isotopes** (hence the common name "**hydrogen bomb**").
- **Energy** from the **fusion explosion** is itself used to trigger **fission reactions** in a **tamper of materials** that would **not otherwise undergo them** – such as the "**depleted**" **uranium** left

behind after the most **fissile isotopes** have been removed.
- By **nesting multiple fusion and tamper layers** together, it's **theoretically possible** to construct **thermonuclear weapons** of **any desired power** – though the **size** of weapons with **more than two stages** is a **challenge to delivery**.
- Although **fusion** itself **does not generate radioactive products**, the **fission** stages still leave behind **large amounts of dangerous fallout**.

LITTLE BOY AND FAT MAN

The bombs detonated over **Hiroshima** and **Nagasaki** at the end of the **Second World War** used a **uranium gun-type trigger mechanism** and a **plutonium implosion** respectively, releasing **energy** equivalent to **15,000** and **21,000 tonnes** of **TNT explosive**.

THE TSAR BOMBA

In October 1961, the **Soviet Union** tested the largest **thermonuclear bomb** ever assembled – a **multi-stage device** known in the West as the **Tsar Bomba** that released **energy** equivalent to **100 million tonnes** of TNT.

THE QUANTUM REVOLUTION

On the smallest subatomic scales, particles are governed not by classical Newtonian physics, but by the curious properties of the quantum universe. The discovery of quantum physics transformed early twentieth-century science.

QUANTIZED LIGHT

In 1900, German physicist **Max Planck** came up with a new attempt to describe the problem of **black-body radiation** (see p.62). **Mathematical laws** attempting to predict the **power output** of perfect radiation emitters at **different wavelengths** tended to work for either **short** or **long wavelengths**, but **not for both**.

Planck's ingenious **mathematical solution** was to imagine that for some reason, black bodies **release their energy** in **small and distinct packets** or "**quanta**" of **light**. Each packet has an **energy** given by the equation

$$E = h\nu \text{ or } E = h(c/\lambda)$$

(where h is a constant – **Planck's constant** – and ν is the **frequency of light**).

Planck's solution brought the equations **back into line with reality**, but he didn't think the quanta had any more significance than as some **curious mode of emitting light** unique to black bodies.

THE PHOTOELECTRIC EFFECT

In 1905, **Albert Einstein** invoked Planck's idea to explain **another puzzling phenomenon**. The **photoelectric effect** is the way in which **electric current flows** through the **surface of certain metals** when they are **illuminated**. But the **relationship** between the **illuminating light** and the production of electrons causing **current to flow** was puzzling.

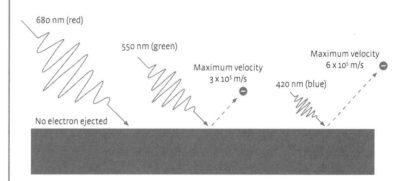

- Red light: No current however intense
- Green light: Current increases with intensity
- Blue light: Current increases with intensity

Einstein showed the effect could be explained if the **electrons** got their **energy** from **individual quanta of light**, rather than a **continuous wave**.

- **Energy** delivered to **atoms** is governed by **Planck's law**.
- **Intensity of light** depends on the **number of quanta striking the surface**.
- **Shorter wavelengths** deliver **more energy**.
- **Longer wavelengths**, however intense, deliver **no more energy** in their **individual quanta**.

Later physicists referred to the **packets** or **quanta of light** as **photons**.

WAVE–PARTICLE DUALITY

Einstein showed that waves of light have discrete particle-like behaviour, but in the 1920s physicists began to question whether the opposite was true – could particles sometimes behave like waves?

DE BROGLIE'S HYPOTHESIS

Einstein's suggestion that **photons were real** implied that they must somehow possess **momentum**, despite lacking **measurable mass**. This momentum p is related to the **frequency** ν and **wavelength** λ by the equation

$$p = h\nu/c \ \text{ or } \ p = h/\lambda$$

(where h is the **Planck constant** and c the **speed of light**).

In his 1924 PhD thesis, **Louis Victor de Broglie** asked why the same equations should not apply to **traditional matter particles**. If a **moving electron**, for instance, has **momentum**, then why shouldn't it also have an **associated wavelength**?

This so-called **de Broglie wavelength** is given by the equation

$$\lambda = (h/mv) * \sqrt{(1-v^2/c^2)}$$

where m is the particle's **mass**, v its **velocity**.

From de Broglie's equation we can tell that:
- The wavelengths of all but the **tiniest particles** are **vanishingly small** (far smaller than the **wavelengths of visible light**).
- **Increasing the speed** of a particle **decreases its wavelength**.
- **Increasing the mass** of a particle also **decreases its wavelength**.

Therefore this **wave-like property** is only apparent among the **smallest subatomic particles**.

THE DOUBLE-SLIT EXPERIMENT FOR PARTICLES

Thomas Young's famous double-slit experiment of 1800 settled the nature of light for a century by demonstrating how it produced **wave-like interference patterns**. In the mid-1920s, **Clint Davisson** and **Lester Germer** showed that **electrons** could produce a **similar effect** when they were **diffracted** by **reflection** on a **nickel surface**.

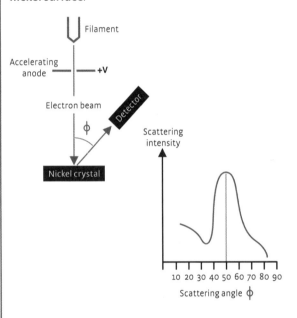

- An **electron gun** (similar to a **cathode-ray tube**) fires a **spreading beam of electrons** towards a **barrier** with **two narrow parallel slits**.
- In practice, the electrons form a **complex interference pattern** – their **wave-like aspect** is **diffracted** in a similar way to **light waves**, creating **ripples** that **interfere** with each other.
- The **interference effect** in the double-slit experiment **only works for the smallest and least massive subatomic particles**.

ELECTRON MICROSCOPY

The wavelengths associated with subatomic particles are far smaller than those of light. Electron microscopes harness this property to create pin-sharp images of objects at magnifications that cannot be achieved with visible light.

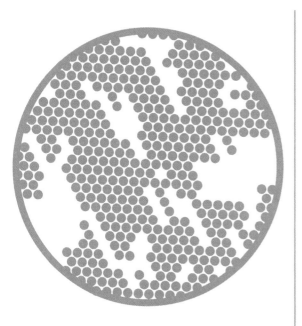

SCANNING ELECTRON MICROSCOPE

The more sophisticated **scanning electron microscope** (SEM) was developed in the 1950s. It forms images by **detecting the way electrons bounce off a reflecting surface**, allowing the **detailed imaging** of **three-dimensional objects** and much **larger samples**.

- Electron beam is **scanned back and forth** across the sample at **high speed**.
- Electrons **bounce back** off the surface, experiencing **scattering** and **diffraction**.
- **Detectors** pick up the **reflected electrons** and use them to **construct an image** of the surface.
- SEMs are limited in **magnification** to about **1 million times**.

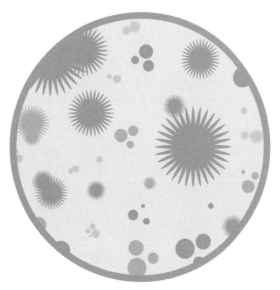

TRANSMISSION ELECTRON MICROSCOPES

The simplest **transmission electron microscopes** (TEMs) were developed in the early 1930s. These rely on passing an **electron beam** through a **very thin sample of material**.

- **Electrons** are fired at the **target** from a **cathode-ray beam**.
- As they pass through the sample, they are affected by **similar phenomena** to those that affect **other waves**, including **scattering** and **diffraction**.
- The electrons are used to produce an **image** on a **phosphor-coated screen**, or are **absorbed** on the surface of **photographic film**, triggering **changes in its chemistry similar to those caused by light**.
- The **resulting image** can have a **magnification** of up to **10 million times**.

THE QUANTUM WAVEFUNCTION

If quantum particles have a wave-like aspect, then it's obviously useful to know the exact shape the waves take. In 1925, Erwin Schrödinger devised an equation that describes just this.

A WAVE EQUATION

Schrödinger coined the term "**wavefunction**" to describe the wave's changing **size**, **shape**, and **strength**, and denoted it using the Greek letter Ψ (psi). He developed several different "**wave equations**" to describe the wavefunction's **properties** and **relationships**, of which the easiest to understand is the one that describes the wavefunction's **evolution over time** in a **single dimension of space**.

(NB: in reality the **wavefunction evolves across a volume of space** in **three dimensions**, giving rise to **far more complex mathematical expressions**.)

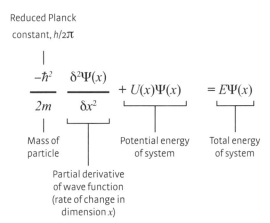

Reduced Planck constant, $h/2\pi$

$$-\frac{\hbar^2}{2m}\frac{\delta^2\Psi(x)}{\delta x^2} + U(x)\Psi(x) = E\Psi(x)$$

Mass of particle

Partial derivative of wave function (rate of change in dimension x)

Potential energy of system

Total energy of system

SO WHAT DOES IT MEAN?

Even today, the **true meaning of the wavefunction** remains a subject of **debate** among physicists, with several **rival interpretations**. For practical purposes, however, we can say that:

- The wavefunction describes the way in which the **properties of a particle** are **spread out across space** according to **experiments** that are designed to **reveal wave-like properties**.
- It **predicts the likelihood** of making a **"classical" observation of the particle** at a **particular location**, or with **particular properties**.

QUANTUM TUNNELLING

Schrödinger's equations underlie much of quantum physics. One of their most important explanations is to explain **what happens during radioactive decay** and **other processes** that should be **impossible** in **classical physics**.

- **Classical picture**: **Nucleus** is surrounded by an **energy barrier** (the "**potential well**") that **prevents particles with limited energies from escaping**.
- **Quantum picture**: **Alpha** or **beta particle** exists *within* **the nucleus**. At a particular moment, the **wave equation** extends *beyond* **the barrier**, giving a **small but finite chance** of the **particle being observed there**.

Classical mechanics

Quantum mechanics

QUANTUM MECHANICS

Quantum mechanics is the name for a particular set of tools used to describe the quantum behaviour of particles. Developed in the mid- to late 1920s, it is also known as wave mechanics or matrix mechanics.

WAVES AND MATRICES

Early in the development of quantum physics, **two distinct approaches** evolved:

- **Wave mechanics** relies on **mathematical manipulations** of the **wavefunction** and **Schrödinger's wave equations**.
- **Matrix mechanics** manipulates **mathematical grids of values** called **matrices** in order to **predict quantum properties**.

$$f_{m,n} = \sqrt{\frac{h}{2\pi}} \begin{bmatrix} f_{11} & f_{12} & f_{13} & f_{14} & f_{15} & ... \\ f_{21} & f_{22} & f_{23} & f_{24} & f_{25} & ... \\ f_{31} & f_{32} & f_{33} & f_{34} & f_{35} & ... \\ f_{41} & f_{42} & f_{43} & f_{44} & f_{45} & ... \\ \vdots & \vdots & \vdots & \vdots & \vdots & \ddots \end{bmatrix}$$

In 1927, **Paul Dirac** developed **"transformation theory"**, which showed that **wave** and **matrix mechanics** are both just **specialized, mathematically equivalent approaches to the same basic problems**.

COLLAPSING THE WAVEFUNCTION

Describing the **behaviour** of actual quantum systems such as **electrons in an atom** means somehow **bridging the gap** between **two very different views of reality**:

- The **wave-like quantum-level description** offered by the **wavefunction**, in which **particles** and their **properties** can occupy a **broad range of positions and energies**, best described in terms of **statistical probabilities**.

- The **large-scale** or **macroscopic view of localized particles**, **precise properties**, and certain **outcomes** that we **experience in the everyday world**.

Physicists often view the wavefunction as a **description of a particle's many possible states** that are "**superposed**" on each other like **interfering waves** of **different wavelengths**. Somehow, these superposed states **always resolve themselves** into a **single**

observed outcome somewhere in the **transition from the quantum to the macroscopic world**.

This **transformation** is known as the **collapse of the wavefunction** – but there are many questions about the **true physical nature of the wavefunction**, and whether it **collapses at all**. These are answered, to a greater or lesser extent, by **various quantum interpretations**.

THE COPENHAGEN INTERPRETATION

The most famous and widely accepted quantum interpretation, the Copenhagen interpretation, attempts to model the collapse of the quantum wavefunction by giving an important role to observers and their measurements.

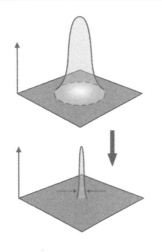

WHY COPENHAGEN?

The **simplest** and **most famous** approach to the **collapse of the wavefunction** is the **Copenhagen interpretation**, developed by **Niels Bohr**, **Werner Heisenberg**, and others from the 1920s onward. Although **never formally defined** by its authors, it is based on a handful of **guiding principles**, of which the most important are:

- The Copenhagen interpretation **disregards questions about the objective reality of the wavefunction**, treating it simply as a **tool for predicting the probability of different outcomes** when the system is **measured classically**.
- An **unobserved quantum system exists** in an **indeterminate state** described by a wavefunction, but at the **moment it is measured or observed** the wavefunction **collapses to a definite outcome**.

THE MEANING OF MEASUREMENT

Followers of the Copenhagen interpretation still **differ** in their views of **what exactly causes the wavefunction to collapse**. **Rival theories** include:

- **Decoherence**: According to this popular model, **macroscopic objects** such as **measuring devices** have **wavefunctions of their own**. When a device **comes into contact** with the quantum system, **interference** between the two of them causes the wavefunction being measured to **lose its stability** or **coherence**, collapsing to a **single definite state**.

- **Spontaneous collapse**: This theory suggests that quantum wavefunctions can **collapse of their own accord**, affecting their surroundings due to **entanglement** (see p.140). The process is **very rare for a single particle**, but a measuring device contains **so many particles** that it is happening all the time. As soon as the quantum system is **brought into contact** (**entangled**) with the measuring device, therefore, it too is affected by the **spontaneous collapse of particles** in the device.
- **"Consciousness causes collapse"**: This **controversial idea** suggests that the **presence of a conscious, thinking observer** is vital to the actual process of wavefunction collapse.

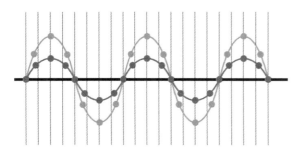

HEISENBERG'S UNCERTAINTY PRINCIPLE

One important result of Heisenberg's "matrix mechanics" approach to quantum theory was to reveal how certain properties are intrinsically linked in measurements. This places inherent limits on our knowledge of the quantum world.

COMPLIMENTARY PROPERTIES

Heisenberg's principle is expressed in the **simple equation**:

$$\Delta x \, \Delta p \geq h$$

where:

Δx is the **uncertainty in position**
Δp is the **uncertainty in momentum**
h is **Planck's constant**

Planck's constant is **tiny** (in standard units its value is 6.626×10^{-34} m² kg/s), so in practice it's **very rarely that we need to measure values of position and momentum** accurately enough running into this limit – and also very difficult to do!

Nevertheless, it's **theoretically impossible to measure the two complimentary measurements with absolute arbitrary accuracy at the same time** – constraining one more precisely leads to **greater uncertainty** in the other.

MEASUREMENT OR REALITY?

1 Uncertainty of position: the more accurately the object's wavelength/momentum is known the less accurately its position can be determined.

2 Uncertainty of wavelength: the more tightly constrained the object's location, the harder it is to pinpoint its wavelength and momentum.

One interpretation of the uncertainty principle is that it's simply a reflection of **limitations on how we measure quantum properties**. **Heisenberg** himself initially looked at the uncertainty principle in this way, but later physicists came to the conclusion that the **measurement issue is irrelevant – uncertainty among complimentary properties** is a **fundamental aspect of quantum behaviour**.

MORE UNCERTAINTY

Although the **link between position and momentum** is the **most famous** aspect of the uncertainty principle, the equations of quantum physics give rise to others. The most important of these is the **time– energy uncertainty relation**:

$$\Delta E \, \Delta t \geq h$$

This states that it is **impossible to pin down the energy of a system** with **arbitrary accuracy** at a **single moment in time**. Therefore energy levels in the system can **fluctuate dramatically**, giving rise to **unexpected and important effects** such as the spontaneous creation of virtual particles.

SCHRÖDINGER'S CAT

Erwin Schrödinger, discoverer of the quantum wavefunction, objected to the reductive approach of the Copenhagen interpretation. To point out what he saw as its absurdities, he invented what is the most famous "thought experiment" in all of physics.

A CAT, A BOX, AND A VIAL OF POISON

Schrödinger's hypothetical experiment involves translating the **inherent uncertainty of the quantum world** onto a **macroscopic scale**:

A **cat is sealed inside a box** so it **cannot be seen** by outside observers.

Inside the box is a small **vial of poison** whose release will **kill the cat**.

The **poison-release mechanism** is only triggered if a Geiger counter detects an **alpha particle** from a small **radioactive source** that is **also sealed inside the box**.

The **radioactive source** is chosen so that, in the course of the experiment, there is a **50/50 chance of a decay event occurring**. If it does, the **poison is released** and the **cat dies**.

ABSURD OUTCOMES

Schrödinger's concern was to point out that the Copenhagen view – that the **wavefunction collapses to a defined outcome at the moment it is observed** (and no sooner) – could lead to **bizarre consequences**. According to Copenhagen, the **state of the radioactive source** remains in a "**quantum superposition**" of **possible outcomes** until the box was **opened**. But surely that means that the **rest of the system** is left in a **similar state of uncertainty**? Until the box is opened, does the cat somehow **hover midway between life and death**?

STRICTLY FOR DEBATE

Despite the lasting fame of Schrödinger's experimental concept, any attempt to actually **carry out the experiment** would be **pointless** as well as cruel. If the wavefunction does indeed **remain in limbo** until it is **observed**, then by definition we **cannot devise an experiment to know this** – whether it's the **opening of the box** or **some other effect** that **causes quantum uncertainty to resolve into concrete reality**, we can never hope to catch the **moment of resolution** – opening the box would always reveal **one outcome or the other**.

MANY WORLDS AND OTHER QUANTUM INTERPRETATIONS

Although the Copenhagen interpretation is still often treated as the standard description of how the wavefunction works, the case is far from settled, and many physicists have come up with alternative approaches.

MANY WORLDS

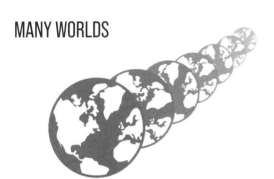

First proposed by **Hugh Everett III** in 1957, the **many worlds interpretation** is perhaps the **boldest** of all in its interpretation of the quantum wavefunction. In many worlds:

- The **wavefunction never collapses**.
- Instead, whenever we "**measure**" a quantum system (in other words, whenever **quantum events interact with the large-scale Universe**), **branching realities** are created for **every possible outcome**.
- Many worlds therefore suggests that the **Universe we observe** is **one among an infinite number of multiverses**.
- The **outcomes of quantum events** are therefore simply a reflection of the fact that **we exist in a particular branch of the multiverse**.

DECOHERENCE

Several alternatives to the Copenhagen interpretation use the concept of "**decoherence**" – the idea that the wavefunction's "**collapse**" is an **illusion**: it can somehow **appear to have collapsed** from a particular point of view, **while in fact remaining intact**.

CONSISTENT HISTORIES

This interpretation uses **complex mathematics** in what is effectively an **expansion of** the **Copenhagen interpretation**. It suggests that the **wavefunction's real purpose** is **not restricted** to **individual quantum events**, but that instead it describes **potential outcomes for an entire system** – a **combination** of **quantum-** and **classical-scale events** that can be as **big as the entire Universe**.

The **consistent histories interpretation** does not claim that **all the different outcomes occur**, or even offer a **tool for predicting which will occur** in a specific system – it is simply a **mathematical means of describing the Universe as we observe it** while **avoiding the question of wavefunction collapse**.

THE ENSEMBLE INTERPRETATION

Favoured by **Einstein**, this view of the quantum world imagines the wavefunction as **describing outcomes across a huge array or ensemble of identical systems** (somewhat similar to **Everett's many worlds**). The wavefunction determines **which of these worlds we find ourselves in** – but again the theory offers **no tools for making predictions**.

QUANTUM NUMBERS AND THE PAULI EXCLUSION PRINCIPLE

In contrast to our everyday experience, many of the properties of subatomic particles are "quantized". Instead of varying continuously, they can only take on discrete values that are represented by "quantum numbers".

QUANTIZED PROPERTIES

Quantum numbers are the **multiplying factors** that describe the **values of quantized subatomic properties**. They may be **multiples of a basic unit** such as **electric charge**, or **unitless numbers** that help make some sort of **distinction**. **Electrons orbiting within atoms**, for example, have **positions** and **energies** described by **four key quantum numbers**:

- Principal quantum number: n
- Azimuthal quantum number: ℓ
- Magnetic quantum number: m_ℓ
- Spin projection quantum number: m_s

PAULI EXCLUSION PRINCIPLE

Wolfgang Pauli's principle simply states that within a particular quantum system such as an **atom**, **no two matter particles** can have the **exact same set of quantum numbers**.

This explains, for instance, why **electrons orbiting an atom** don't simply fall towards the **lowest possible energy state** close to the **nucleus**, but instead occupy a **complex pattern of orbital shells**.

THE QUANTUM ATOM

The four electron quantum numbers define the **structure of orbital shells** around an **atom**:

- The **principal quantum number** n defines the **overall shell**.
- The **azimuthal quantum number** ℓ defines the "**subshell**".
- The **magnetic quantum number** m_ℓ defines the specific **electron "orbital"**.

As n **increases**, so too does the range of **possible values of l**, and as l **increases** so does the possible range of **values for m_ℓ**. The **spin quantum number** m_s, however, remains **unchanged** with only **two possible values**.

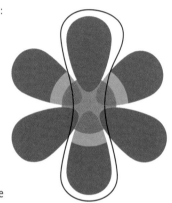

Relationships among values of n, l, and m_ℓ through n=4

n	Possible values of l	Subshell designation	Possible values of m_ρ	Number of orbitals in a subshell
1	0	1s	0	1
2	0	2s	0	1
	1	2p	1, 0, -1	3
3	0	3s	0	1
	1	3p	1, 0, -1	3
	2	3d	2, 1, 0, -1, -2	5
4	0	4s	0	1
	1	4p	1, 0, -1	3
	2	4d	2, 1, 0, -1, -2	5
	3	4f	3, 2, 1, 0, -1, -2, -3	7

The result is an **increasing range of electron orbitals** in an increasing number of **subshells** further from the **nucleus**, as seen in the **periodic table**. Each **orbital** can be occupied by just **two electrons** with different **spin values**.

SPIN

Spin is the subatomic quantum equivalent of a particle's angular momentum in classical mechanics. It has strange and useful properties, and helps define a fundamental division among particles.

WHAT IS SPIN?

Spin is commonly depicted as **rotation of a subatomic particle about its axis** in either a **clockwise** or **anticlockwise** sense. This analogy only goes so far, however. In reality **spin is rather different**: it isn't caused by **physical rotation** in the **classical sense**; it is "**quantized**" so that particles can only have **certain amounts of spin** (rather than **varying continuously**) and it **accumulates** rather more like **electric charge** than **angular momentum**.

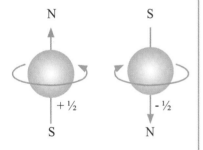

Spin is classed as **positive** or **negative** depending on its relationship to the **particle's magnetic field**. This plays an important role in **atomic structure**, since the **two possible spin values of electrons** (+ ½ and - ½) permit a **pair** of electrons to **share the same atomic orbital** without breaking the **Pauli exclusion principle**.

FERMIONS AND BOSONS

Different **values of spin** help distinguish between **two fundamental families of particles in nature**:

- Particles with **half-integer spins** (such as **electrons**) are known as **fermions**, while particles with **whole-number** or **zero spin** are called **bosons**.
- Only **fermions** are affected by the **Pauli exclusion principle**, so they **behave in a very different way from bosons**.
- Among the **elementary (indivisible) particles of nature**, all the particles **known to make up matter** are **fermions**, while **bosons** are restricted to **massless "messenger particles"** used for **transmitting forces**.

SPIN AT WORK

Magnetic resonance imaging (MRI) is a hugely important medical technique used for studying **soft tissues in the human body**. **Radio waves** and **magnetic fields** are used to **align the spin of protons (hydrogen nuclei**, plentiful in the body's water content) in the **same direction**. When the nuclei are allowed to **return to their original spin directions**, they **emit radio waves**, which can be used to **map the body's organs**. The **time** that nuclei in various areas take to **relax** is linked to the **properties** of **different tissues**, providing a powerful **diagnostic tool**.

SUPERFLUIDS AND SUPERCONDUCTORS

The Pauli exclusion principle applies to all fermions (elementary particles of matter) and is hugely important to the structure of matter itself. On the rare occasions when it ceases to apply, strange phenomena arise.

SUPERFLUIDS

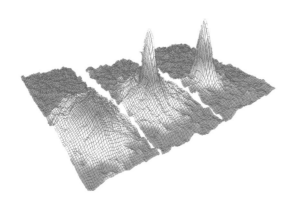

Although all **elementary matter particles** are **fermions**, there is another class of particles – **bosons** – that are **immune to the exclusion principle**. These have **whole-number** or **zero spin**, and **obey a different set of quantum rules** called **Bose–Einstein statistics**.

While the **only single-particle bosons** that **exist in nature** are **massless force-transmitters** such as **photons**, **composite bosons** made from **matter particles** are also possible.

Composite bosons form when **even numbers of fermions bind together**: spin **adds up** rather like **electric charge**, so **linking two fermions** with **equal** or **opposite spins** results in a **net spin** of 1 or 0 respectively.

$$\text{Spin} + \tfrac{1}{2} + \text{spin} - \tfrac{1}{2} = \text{Spin } 0$$
$$\text{Spin} + \tfrac{1}{2} + \text{spin} + \tfrac{1}{2} = \text{Spin } 1$$

The same principle applies **however many pairs of fermions are linked**, so **atoms** such as **helium-4** (with **two neutrons**, **two protons**, and **two electrons**, all of which are **fermions**) can **behave as bosons**.

- At **normal temperatures**, particles made of **composite bosons** have enough **energy** to keep them in a **range of different states**. But when they are **cooled below a critical level**, all the atoms can sink into a **new state of matter** called a **Bose–Einstein condensate** (BEC).
- BECs act as if they are a **single enormous particle**, and exhibit **strange behaviours**, including a **complete lack of internal friction** that allows them to **move very quickly** as "**superfluids**".

SUPERCONDUCTORS

When certain materials are **cooled to very low temperatures**, the **electrons** can **come together** to form **weakly bonded "Cooper pairs"** that also act as **composite bosons**. As with **superfluids**, the effect of sharing **identical quantum properties** reduces **interaction with their surroundings** and allows the **electrons** to **flow in a frictionless manner**, meeting **no electrical resistance** and becoming a **super-efficient electrical "superconductor"**.

QUANTUM DEGENERACY

The Pauli exclusion principle is remarkably powerful. In some extreme situations, it becomes the only thing that prevents matter from collapsing completely to its lowest possible energy state.

DEGENERATE MATTER

Degeneracy occurs when matter is **compressed** into an **extremely small space**. High density **concentrates matter** and **confines its position**, which (as a consequence of the **uncertainty principle**) means that its **momentum** and **kinetic energy** become **increasingly ill-defined**. The result is that within a **tiny range of movement, particles jostle** each other at **very high speed**, producing a "**degeneracy pressure**" that **resists further compression**.

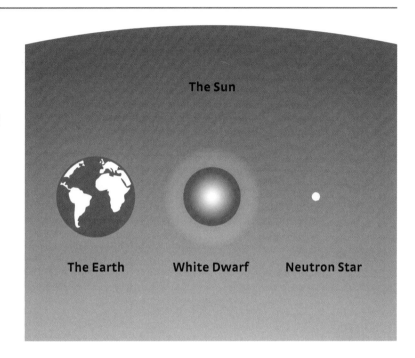

The Sun

The Earth · White Dwarf · Neutron Star

DEGENERATE STARS

Degenerate matter occurs most commonly in the **cores of dead stars**. When a star dies, the **outward pressure from its radiation stops** and the core begins to **fall inwards** under the **pull of its own gravity**. The collapse may be **slow and steady** in the case of stars like the **Sun**, or **near-instantaneous** and **extremely violent** in the case of **more massive stars**.

- In stars like the Sun, the **collapse halts** when **free-floating electrons** in the **core** become **degenerate**. By this time, it may have collapsed to about the **size of the Earth**, becoming an **extremely hot, slowly cooling white dwarf star**.
- In a **high-mass star**, the **final exhaustion of the core** leads to a **sudden inward collapse** and a **shockwave**

that is **powerful enough to overcome electron degeneracy pressure**. To preserve the **exclusion principle**, **electrons** are forced to **merge together** with **protons**, forming **uncharged neutrons** that cram into a **much smaller space**.

- As the core reaches a **few kilometres across**, the **degeneracy pressure** between **neutrons** finally **brings its collapse to a halt**, leaving a **city-sized stellar remnant** called a **neutron star**.
- It's theoretically possible for a star's collapse to be **so violent** that it **overcomes even neutron degeneracy**. The particles then **disintegrate** into their **constituent quarks**, which can **exert their own degeneracy pressure** at an **even smaller size** – just a few kilometres across.

ENTANGLEMENT

The effects of quantum uncertainty don't just apply to single particles – they can also extend across systems of related, interdependent objects. This gives rise to one of the strangest of all quantum effects – something Einstein himself called "spooky action at a distance".

BOUND TOGETHER

Entanglement is an **intrinsic link between the states of microscopic particles** that **allows them to share information instantaneously**. No matter **how widely the particles are separated**, they remain **bound together by entanglement**, so that, in effect, **information can cross between them faster than the speed of light**.

Entanglement arises when a **pair of subatomic particles** is **put through a special procedure** that ensures their **quantum properties** must be **related to each other**:

- A classic example is to **force a pair of electrons towards the same quantum state** – in order to obey the **Pauli exclusion principle**, they will have to take on **opposite spins** of $+\frac{1}{2}$ and $-\frac{1}{2}$.
- The relationship can be established **without having to measure either particle's spin** directly, so their **wavefunctions remain uncollapsed**.
- When the **spin of one particle** is **eventually measured**, the **wavefunction of the other** will **instantly collapse** to the **opposite state without a signal having to travel between them**.

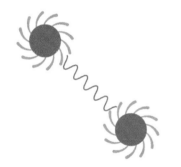

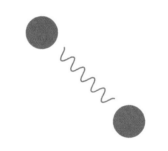

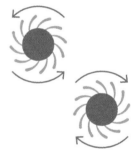

QUANTUM TELEPORTATION

Although it's light-years away from the kind of large-scale **teleportation** seen in *Star Trek*, physicists have successfully used **entanglement** to create **perfect replicas** of **photons**, **subatomic particles**, and even **entire atoms**. The process involves "**scanning**" the subject particle using **one half of an entangled quantum particle pair**. This **modifies** its **entangled partner** in ways that allow a **duplicate** of the original subject to be created. In theory, it **would be possible to teleport objects instantaneously over huge distances** – the only catch is that the **original subject is seriously disrupted** in the process.

QUANTUM COMPUTING

The dream of building a computer capable of solving seemingly impossible problems using quantum physics has only recently become a reality.

PROCESSING WITH QUBITS

A YOUNG TECHNOLOGY
While quantum computing has **enormous theoretical potential**, there are huge **challenges** involved in **putting it into practice**.

A qubit itself may be **any particle capable of being put into a quantum superposition** (including **atoms, ions, photons**, and **individual electrons**), but it must somehow be kept in **complete isolation** from the **surrounding system** in order to **prevent its wave function "decohering"**.

As a result, **progress** towards quantum computing has been **slow**. True systems with multiple qubits remain hugely challenging, and **claims of up to 2,000 qubits** are **confined to highly specialized devices** rather than computers with general applications.

Quantum computers are devices that **manipulate data** by **using a variety of quantum effects**. Specifically, they **store data in qubits** (the **quantum equivalent** of the **binary bit**) – **particles** that are **held in a superposition of all their possible states**, described by a **wavefunction**.

A qubit can **represent all possible states simultaneously**. For a single qubit this may be a **simple choice** between a **digital 0 or 1**, but the **ability to represent both values at the same time** means that the **number of possible states increases exponentially** as **more qubits are coupled together**:

$$1 \text{ qubit} = 2 \text{ states}$$
$$2 \text{ qubits} = 4 \text{ states}$$
$$3 \text{ qubits} = 8 \text{ states}$$
$$n \text{ qubits} = 2^n \text{ states}$$

So, for example, a system of 64 coupled qubits can represent 1.84×10^{19} possible states. A **measurement process** causes the system's **combined wavefunction** to **instantaneously collapse** and **reveal a solution**, making quantum computers a **potentially powerful way of solving "brute-force" problems with large numbers of possible solutions**.

THE PARTICLE ZOO

Particle physics is the branch of physics concerned with the fundamental building blocks of the Universe – the elementary particles that make up matter, and the forces that bind them together.

ELEMENTARY PARTICLES

According to current thinking, an **elementary particle** is one that **cannot be split further into smaller particles**:

- The **negatively charged electrons** found in **all atoms** are elementary.
- The **protons** and **neutrons** of the **atomic nucleus** are **not elementary**, since they are **each made of three smaller particles** called **quarks**.
- **Photons**, or **single-wave packets of light**, are also elementary.

TYPES OF PARTICLE

Matter particles can be **distinguished from each other** in various ways, depending on whether they are influenced by different **fundamental forces**:

Particles with **mass** are affected by the **gravitational force**.

Particles with **electric charge** experience the **electromagnetic force**.

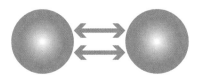

Quarks are affected by the **strong** and **weak nuclear forces**.

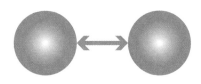

Leptons (such as the **electron**) are affected *only* by the **weak nuclear force**.

The elementary particles from which **matter** is made (including **quarks** and **leptons**) are known as **fermions**, while the particles responsible for **transferring the forces** between them are called **bosons**.

ANTIMATTER

Antimatter is simply **matter made up of elementary particles** with the **opposite electric charge to normal matter**.

Antimatter particles are **rare** in the Universe because they **annihilate on contact with normal matter**, disappearing in a **burst of energy** (usually emitted as **gamma rays**).

THE STANDARD MODEL

The Standard Model of particle physics describes the different elementary particles that scientists agree must exist. However, it does not resolve every question in physics – other elementary particles may still await discovery.

STANDARD TABLE

The **Standard Model** divides particles into **fermions** and **bosons** depending on their **spin** – the **quantum property** that determines **how particles behave in confined systems** (see p. 137). Fermions are **particles with spin ½** (in **either direction**), while **bosons** have **whole-number or zero spins**.

(see p. 137)

- There are six **quarks**, paired together in three **"generations"** of broadly **increasing mass and energy**. One quark in each **pair** has an **electric charge** of +2/3, the other a charge of -1/3.
- Six **leptons** are similarly grouped in **pairs**, consisting of a particle with an **electric charge** of -1, and its **electrically neutral counterpart**, called a **neutrino**.

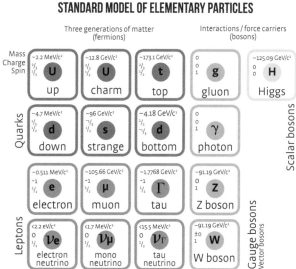

STANDARD MODEL OF ELEMENTARY PARTICLES

Three generations of matter (fermions)

Interactions / force carriers (bosons)

- The **bosons**, meanwhile, consist of **five particles** with spin 1:
  Photon (carries electromagnetic force)
  Gluon (transmits strong nuclear force)
  W⁺ and W⁻ (**electrically charged carrier** of the **weak nuclear force**)
  Z⁰ (**neutral carrier** of the **weak nuclear force**)
  ... and a sixth with spin 0 – the famous **Higgs boson**.

PARTICLE PHYSICS

PARTICLE MASSES

Scientists measure the **tiny masses of elementary particles** using their **energy equivalent**, according to Einstein's famous equation $E=mc^2$.

The units of energy involved are **electronvolts** (eV) – the **amount of energy gained or lost when a single electron moves through a potential difference of 1 volt in a vacuum**.

$$1 \text{ eV} = 1.602 \times 10^{-19} \text{ joules}$$

Turning Einstein's equation around gives the solution $m = E/c^2$.

Particle masses are therefore expressed in terms of **electronvolts/c^2**. In practice, most particles in the **Standard Model** have masses **millions or billions of times larger** than this, written **MeV/c^2** or **GeV/c^2**.

QUARKS

Quarks are the elementary particles found in heavy subatomic particles such as protons and neutrons. They are never seen in isolation, but their existence has been proven in a variety of experiments.

QUARK GENERATIONS

Experiments have identified **three generations of quarks** in **six different "flavours"**: **up/down**, **strange/charm** and **top/bottom**. Only the **up and down quarks** are found in the **everyday matter of today's Universe** – the others can only **come into existence briefly** when **vast amounts of energy are released in particle accelerators**.

Quarks of charge $+2/3$: **up**, **charm**, **top**
Quarks of charge $-1/3$: **down**, **strange**, **bottom**

COMBINING QUARKS

Quarks can combine together in **pairs**, **triplets**, or **larger groups**. Particles formed by **combined quarks** are known as **hadrons**, but are subdivided into **mesons** and **baryons**:

- **Mesons** contain an **even number of quarks** (usually two, consisting of a **quark–antiquark pair**). They are **extremely short-lived**, **decaying** to produce other particles before they can **annihilate**.

The **most common mesons** are the three **pions**:
π^+ = up + antidown

π^- = down + antiup

π^0 = up + antiup or down + antidown.

- **Baryons** contain an **odd number of quarks** (three or more). They include the **everyday matter particles** the **proton** (two **up quarks**, one **down quark**) and **neutron** (two **down quarks**, one **up quark**).

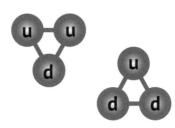

BINDING QUARKS

The **most important force binding quarks** together is the **strong nuclear force**, an extremely powerful force that works only at **very close range**.

Inside a **baryon** or **meson**, the **quarks** are **held together** by the **exchange of particles** called **gluons**.

Baryons and **mesons** themselves bind together **more weakly** by the **exchange of composite meson particles**. The familiar **proton** and **neutron** are held together by **exchange of pions**.

NAMING QUARKS
George Zweig and **Murray Gell-Man** independently proposed the existence of quarks in 1964 as a solution to the **classification of heavier hadron particles**. At first, **only three quarks** were required to **fit observations**, and **Gell-Mann's name for them** came from a nonsense phrase in **James Joyce's** *Finnegan's Wake*: "Three quarks for muster mark".

LEPTONS

Leptons are relatively lightweight fermion particles that are immune to the strong nuclear force. Like quarks, there are three paired "generations" of leptons – but within each generation, the two partner particles are very different.

ELECTRONS AND THEIR ALLIES

The most familiar **lepton particles** are **electrons** (denoted e⁻) – fermions with a **mass** 1/1836ᵗʰ that of a **proton** and a **negative electric charge** that **orbit** in the **outer shells of atoms**.

In **high-energy conditions, negatively charged particles** with **higher mass**, called **muons** (μ^-) and **tau leptons** (τ^-), can **act in a similar way to** electrons.

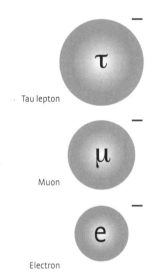

Tau lepton

Muon

Electron

Tau neutrino

Muon neutrino

Electron neutrino

The **electron, muon,** and **tau lepton** are each **paired** with particles called **neutrinos**, which have much **lower masses** and carry **no electric charge**. These particles **pass straight through most forms of matter without interaction**. They are denoted by the Greek letter ν (*nu*) and a subscript.

Lepton interactions frequently involve the **loss or gain of electric charge** (in the form of a **negative electron** or its **positively charged antimatter equivalent**, the **positron**). Loss or gain of neutrinos can help to **balance out the system's overall spin**.

HUNTING NEUTRINOS

Neutrinos are **produced in huge numbers** by the **nuclear fusion reactions** that power the **Sun** (see p.124). **Fusing two protons (hydrogen nuclei)** together forms a **deuterium nucleus**, with **one proton turning into a neutron, shedding charge** and **spin**:

$$p^+ + p^+ \rightarrow {}^2H\ (p^+ + n^0) + e^+\ (\text{positron}) + \nu_e\ (\text{electron neutrino})$$

Countless solar neutrinos pass straight through the Earth every second, and astronomers study them using huge **Cherenkov detectors** (see p.118) located **deep below ground** to **shield them** from other particles. However, **neutrinos oscillate from one form to another** during their **journey from the Sun**, and only **electron neutrinos** can be **directly measured** with these instruments.

DETECTING PARTICLES

Physicists use various methods to detect subatomic and elementary particles, whether they originate in nature or are manufactured in particle accelerators.

EARLY DETECTORS

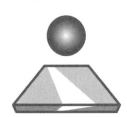

- **Photographic film**: certain particles trigger **chemical reactions** in a similar way to various forms of **electromagnetic radiation**, causing the **film to darken**. **Victor Franz Hess** used stacks of **photographic plates** to discover **cosmic rays** (particles from space) in 1911–13.

- **Cloud chamber**: as particles pass through a **sealed chamber containing air** that is **supersaturated with water vapour**, they trigger the **formation of cloud trails**. A **magnetic field around the chamber** can be used to measure the **polarity of a particle's electric charge**, and its **charge/mass ratio**.

- **Bubble chamber**: in these devices, the **passage of particles** through **transparent liquid** that is **superheated above its boiling point** (but not able to naturally boil) leaves a **trail of bubbles**.

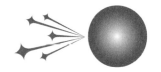

- **Spark detectors**: these devices work rather like **Geiger counters**, with particles **ionizing a gas** and **triggering electric sparks** as they pass through them.

MODERN DETECTORS

- **Drift tubes**: these more sophisticated variations on the **spark detector principle** use a **grid** of **multiple, narrowly spaced wires** to track the passage of particles in **three dimensions**.

- **Electronic detectors**: these use **solid-state silicon circuits** (analogous to **camera CCDs**) that are **wrapped in layers** around the **collision chamber** and record the **passage of particles passing through them**.

- **Cherenkov detectors**: these devices use a medium with a **high refractive index** and **slow internal speed of light**. Particles passing through the medium **exceed its speed of light** and produce **Cherenkov radiation** in **flashes** that are **captured by arrays of light detectors** surrounding the medium.

PARTICLE ACCELERATORS

Particle accelerators are machines for studying the properties of subatomic matter, including elementary particles. They include the biggest machine on Earth – the Large Hadron Collider (LHC).

ACCELERATOR PRINCIPLES

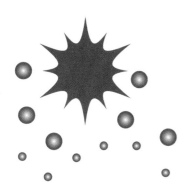

Despite the popular name "**atom smashes**", particle accelerators **do not simply split atoms apart** to see what's inside them:

- **Charged particles** are **boosted to high speed** by **powerful electromagnetic fields**.
- Particles are **smashed into a target** or **each other** with **so much force** that they **transform into pure energy** in accordance with **Einstein's** $E=mc^2$.
- The **energy rapidly condenses** back into particles that have **higher energies** and **masses** than **those that exist** in the **normal low-energy Universe**.
- These **unstable particles** rapidly **decay** into **more familiar forms**.

THE LARGE HADRON COLLIDER

- Location: France/Switzerland border
- Circumference: 27 km
- Depth: Up to 175 metres
- Collision energy: Up to thirteen teraelectronvolts for protons, up to 574 TeV for lead ions
- Main detectors: CMS; ATLAS; ALICE; LHCb
- Operational: 2010

NOTABLE ACCELERATORS

1927 Rolf Wideroe builds the first linear accelerator at Aachen University, using high-voltage electric fields to speed up charged particles to kinetic energies of 50 keV

1931 Ernest Lawrence at the University of California builds the first spiral accelerator or cyclotron, boosting protons to 1.1 MeV

1953 Brookhaven National Laboratory's 72-metre Cosmotron accelerates protons to energies of up to 3.3 GeV

1966 Stanford Linear Accelerator (SLAC); 3-kilometre tunnel accelerating electrons and positrons to 50 GeV

1983 Tevatron; circular accelerator built at Fermilab in Illinois, uses superconducting magnets to boost protons to 980 GeV

2008 Large Hadron Collider; circular 27-kilometre ring accelerating protons up to 6.5 TeV (6,500 GeV)

FUNDAMENTAL FORCES

Four fundamental forces of nature control all the interactions of matter. While gravitation seems to obey its own set of rules, the electromagnetic, weak, and strong forces all share some common features that suggest they work in similar ways.

FOUR FORCES

Each of the fundamental forces has its own characteristic properties:

STRONG NUCLEAR FORCE

Affects only quarks
Effective range: 10^{-15} m
(diameter of a mid-sized
atomic nucleus)
Strength*: 1

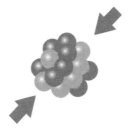

WEAK NUCLEAR FORCE

Affects all fermions
Effective range: 10^{-18} m
(1/1000th the diameter of
a proton)
Strength*: 1/1000,000 (at
10^{-15} m)

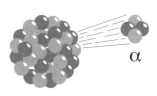

ELECTROMAGNETIC FORCE

Affects all particles with
electric charge
Effective range: Infinite
Strength*: 1/137

GRAVITATION

Affects all particles with
mass
Effective range: Infinite
Strength*: 6×10^{-39} (only
builds up when large
masses accumulate)

(*For simple comparison, strengths are given at the range of the strong nuclear force.)

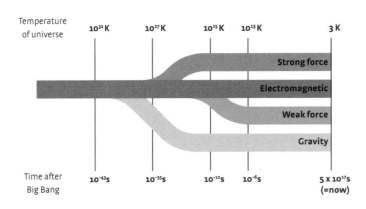

COMMON ORIGIN

Theoretical physicists believe that the **four forces** were originally **"unified"**, behaving as a **single force** in the **high-energy environment** of the **Big Bang**. As the **young Universe cooled**, they **split apart** – **similarities** between the forces today gives **clues as to the order in which this happened**.

GAUGE THEORIES AND QED

How are forces transmitted between elementary particles? Since the 1950s, physicists have developed a number of "gauge theories" to explain the three non-gravitational forces.

GAUGE BOSONS

Gauge theories describe **interactions** between **fermions (matter particles with half-integer spins)** in terms of an **exchange of bosons** with spin 1.

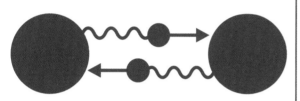

But **where do the bosons come from**? Importantly, they **do not have to be detectable in normal situations** – they can be "**virtual particles**" that live and die on **extremely short timescales** before they can be detected.

Virtual particles are possible because of the **time-energy uncertainty principle** (see p.133):

$$\Delta E \times \Delta t < \hbar / 2$$

This allows **small amounts of energy** to be "**borrowed**" for **brief periods** of time to **create particles**, provided they are "**paid back**" later.

QED

Quantum electrodynamics (QED) was the **first fully developed field theory of a quantum force**. It describes **electromagnetic interactions** in terms of the **exchange of virtual photons**.

Richard Feynman devised a way of **portraying interactions** in **simple diagrams** using **straight lines** to indicate **fermions** and **wiggly lines** to indicate **bosons**:

This simple Feynman diagram shows an electron and a positron annihilating to release gamma rays.

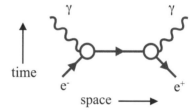

(NB: antiparticles are shown with arrows pointing in the opposite direction to normal particles – so both particles here are approaching their point of interaction or vertex.)

In Feynman's approach, the different possible **interactions** between **charged particles** and **photons** in a system are **mapped out** and the **different probabilities of their happening** assessed.

QUANTUM FIELD THEORIES

Force	Gauge boson	Gauge theory
Electromagnetism	Photon	Quantum electrodynamics
Strong nuclear force	Gluon (and composite pion)	Quantum chromodynamics
Weak nuclear force	$W^{\pm}$, Z^{0} particles	Quantum flavordynamics

STRONG NUCLEAR FORCE

The strong nuclear force, as its name suggests, is the strongest of all the fundamental forces, despite having a very limited range. It is explained by a gauge theory known as quantum chromodynamics or QCD.

ONE FORCE, TWO EFFECTS

The strong force works on two different scales:

- Inside individual **nucleons** (**protons** and **neutrons**) the **strong force** is at its **most powerful**, **binding quarks** together through the **exchange of virtual particles** called **gluons**.
- Some of the strong force's effects "**leak out**" of the **nucleons**, **binding them** together in turn (albeit **more weakly**) through the **exchange of virtual pions** (**quark–antiquark pairs** – see p.144).

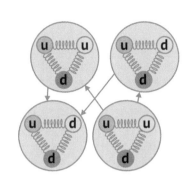

QUARK COLOURS

Quarks are susceptible to the **strong force** because of a **unique property** known as **colour charge**. This has nothing to do with either **visible colour** *or* **electric charge** – but both are **useful analogies** for **modelling quark behaviour**.

- **Quarks** may have "colours" of **red**, **green**, or **blue**, while **antiquarks** have colours **antired**, **antigreen**, and **antiblue**. They form **combinations** that always

THREE-QUARK BARYONS:

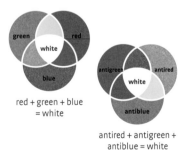

red + green + blue
= white

antired + antigreen +
antiblue = white

TWO-QUARK MESONS:

red + antired
= white

green + antigreen
= white

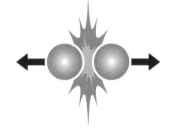

blue + antiblue
= white

balance to appear "**white**" from the outside, **interacting** through the **exchange of gluons**.

- Despite their **overall balance**, the **distribution of quarks** of **different colours** inside **baryons** means that some colour "**leaks out**" in a **weakened form**, allowing **baryons to interact with others close by**. This **"residual" strong force** is transmitted through the **exchange of virtual pions**.

COLOUR CONFINEMENT

The **colour of quarks** is **impossible to observe directly** due to an effect called **colour confinement**. **Breaking one quark away from another** requires **so much energy** that a new **quark–antiquark pair** is **spontaneously created** and instantaneously **binds with any separated fragments** – **breaking a hadron apart** always produces **two new hadrons** rather than a **lone quark**.

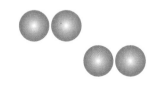

WEAK NUCLEAR FORCE

The weak nuclear force has the shortest range of all. It's also the hardest to understand since it does not merely bind particles together, but allows them to be transformed.

CHANGING FLAVOURS

Uniquely, the **gauge theory** of the **weak force** relies on **not one**, but **three gauge bosons** – the **neutral Z°**, and the **electrically charged W⁺ and W⁻**.

All **elementary particles** (both **quarks** and **leptons**) are subject to the **weak force**, but their **degree of susceptibility** is governed by a quantum property called **weak isospin** (often denoted T_3).

Weak interactions come in two different varieties, called **neutral-current** and **charged-current**:

- **Neutral-current interactions** work in the same way as those of other forces, and involve the **exchange of Z° bosons**.
- **Charged-current interactions** involve the **W⁺ or W⁻ bosons**, and **change the flavour** of a **quark** or **lepton**. The **charged bosons** act in the same way as **elements of a mathematical equation, adding** or **removing electric charge** as they **transform the particle into its partner of the same generation**.

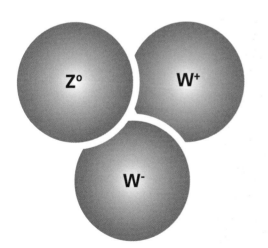

BETA DECAY

- **Radioactive beta decay** (see p.120) involves **one neutron** in an **unstable atomic nucleus** spontaneously **transforming into a proton** and **emitting a beta particle** (electron).

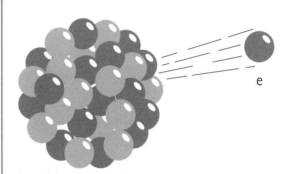

On a **quark level**, changing a **neutron** (up-down-down) into a **proton** (up-up-down) involves transforming a **down quark** into an **up quark**. Emission of a **W⁻ particle** changes the quark's **flavour** and "**carries away**" a unit of **electric charge**:

$$d \ (\text{charge } -^1/_3) \rightarrow u \ (\text{charge } ^2/_3) + W^-$$

The W- particle is itself **highly unstable**, so it **rapidly decays, shedding its energy** to create **two stable leptons**:

$$W^- \rightarrow e^- + \nu_e \ (\text{an electron antineutrino})$$

THE HIGGS BOSON

The Higgs boson is the most famous particle in modern physics, and was the last piece of the Standard Model to be confirmed. It explains why the bosons involved in the weak nuclear force – and the fermions that make up matter – have mass.

THE HIGGS FIELD

Gauge theories predict that **force-transmitting gauge bosons should be massless** – so **why do the W⁺, W⁻, and Z⁰ particles that carry the weak force have substantial mass?**

In 1964, **Peter Higgs** and others proposed the "**Higgs mechanism**" – a method for particles to **acquire mass** through i**nteractions with a field** permeating all of space.

Uniquely, the **Higgs field** requires **less energy** to take on a **non-zero value** than it does for a **zero value**, so it tends to be **non-zero** in even the **lowest-energy conditions**.

Imagine two balls of the **same mass** but different **diameters** falling through a **viscous fluid** such as oil: the ball with the **smaller cross-section** experiences **less friction** from its surroundings and **falls faster**.

DISCOVERING THE HIGGS BOSON

With a **spin of 0, no colour charge** and **no electric charge**, the Higgs is not a "**gauge boson**" in the same sense as the **other elementary bosons** – instead it **arises only** when the **Higgs field** is **boosted into an excited state** – something that only happens during **phase transitions of exotic forms of matter**.

Finding the Higgs was one of the main goals of the **Large Hadron Collider** when it became operational in 2009. A particle with the **correct mass** and other properties was detected during the **first experimental run**, and announced on 4 July 2012.

Confirmed mass of the Higgs boson:

$$125.18 \pm 0.16 \ \text{GeV/c}^2$$

NOT THE WHOLE STORY

The **Higgs mechanism alone** is not enough to account for the **mass** of the **many different fermion particles**. Physicists are still exploring ways for **single fermions to gain additional mass**, while in **composite matter particles**, **binding energy** (see p.123) also plays an important part.

SYMMETRY

In order to understand the mysteries of particle physics, scientists extend the geometrical concept of symmetry to a range of other phenomena, including the properties of elementary particles and interactions involving fundamental forces.

WHAT IS SYMMETRY?

In **geometry**, a figure is symmetrical if it **remains the same after being subjected to a geometric transformation**. The most familiar form of symmetry involves **reflection around a particular axis** to create a "**mirror image**", but other forms or geometric symmetry apply to **rotation** and **translation through space**:

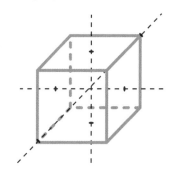

Particles and **force interactions** can also be "**symmetric**" in a sense if they **remain the same after a uniform change is imposed on the particles involved**.

IMPORTANT PARTICLE SYMMETRIES

- **Charge** (C-symmetry): The **electric charges** of **all particles involved** are **reversed** (i.e. each particle is replaced with its **antiparticle**) and the **interaction remains the same**.
- **Parity** (P-symmetry): The **orientations** and **spins** of the particles are **flipped** but the **interaction remains the same**.
- **Time** (T-symmetry): The **flow of time** is **reversed** but the **interaction remains the same**.

Symmetries can also work **in combination**:

- **CP symmetry** involves **reversing the charge** *and* **parity** of all particles involved. This applies to **strong-force** and **electromagnetic interactions**, but is **broken by the weak force**.
- **CPT symmetry** involves **reversing charge**, **parity** *and* **time** – the **Standard Model** insists that this is **true in all circumstances**.

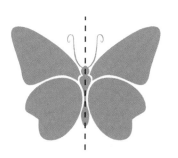

SYMMETRIC FORCES

Physicists believe that in the **instant of the Big Bang** itself, **all four forces were symmetric, acting as one** and **producing the same results**. The **breaking of symmetry** released **huge amounts of energy** that drove the sudden **expansion of the early Universe** in a process called **Inflation** (see p.168).

THEORIES OF EVERYTHING

Is it possible to simplify the four fundamental forces and multiple elementary particles of the Standard Model? Most theoretical physicists believe that it is, creating a Grand Unified Theory, and perhaps even a Theory of Everything.

UNIFYING THEORIES

In extremely **high-energy environments** such as those in **particle accelerators, forces themselves begin to become symmetric**. The **weak** and **electromagnetic forces** produce the **same results**, meaning they can be **unified by a single "electroweak" model**.

Theoretical physicists hope that the **strong force** can be shown to **converge** with the **electroweak force** at even **higher energies** – if so, the resulting **"electronuclear" force** could be described in a so-called **Grand Unified Theory** or **GUT**. Various candidate GUTs predict:

- **Massive new particles** at **energies** of around 10^{16} GeV/c^2 – **far beyond those of any particle accelerator**.

- **Proton decay** – the occasional **spontaneous decay of protons (stable** according to the **Standard Model)** into other particles.

- **Magnetic monopoles – hypothetical** particles with a **magnetic field** but just one **magnetic pole**.

However, **none of these have yet been observed** – so while a GUT is probably possible, we don't yet know **which of many alternative models is correct**.

THEORIES OF EVERYTHING

Unifying gravitation with the **other three forces** in a single "**Theory of Everything**" is an even bigger challenge. While **quantum field theories** can describe the other three forces, **gravitation** is so far **modelled best by general relativity**.

Most theories of "**quantum gravity**" that could **conceivably be unified** with the other forces require the existence of **gravitons – gauge bosons** for carrying gravitational **force**. The graviton's **required properties** are:

- zero mass
- moves at the speed of light
- spin 2

If the graviton exists, its **effects** only **become clear** at the **tiniest scales imaginable** – around 10^{-35} metres. On **bigger scales** than this, it must give rise to the well-known effects of **general relativity** such as the **distortion of spacetime** (see p.160).

STRING THEORIES AND EXTRA DIMENSIONS

Could the various quantized properties of elementary particles be described by an underlying reality in which they are strings of energy, vibrating in many different dimensions?

PARTICLES IN HARMONY?

Open strings   **Closed strings**

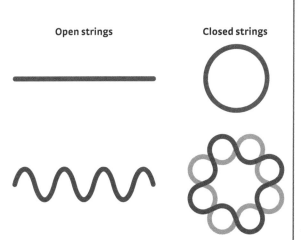

String theory is one of the most popular candidates for a **potential Theory of Everything, uniting all of nature's forces and particles in a single model**:

- Particles are **vibrating strings of energy** with lengths around 10^{-35} m.
- As with any vibrating string, they form **standing waves** with **various harmonic frequencies** and **wavelengths**.
- These **harmonics** determine the **values** of the **various measured particle properties**. Because the string **cannot occupy states between them**, they explain why many of these properties are **quantized in discrete units**.
- However, in order to produce the **range of properties** displayed by particles, the **strings must vibrate in more than the familiar three space dimensions**.

HIGHER DIMENSIONS

Different string theories require **different numbers of extra dimensions**:

- **Bosonic string theory**, developed in the 1960s, requires a total of **twenty-six dimensions** of **spacetime** just to produce **bosons**.
- **Supersymmetric string theory** reduces the number of **spacetime dimensions** to **ten**. It requires each **Standard-Model fermion** to have a **high-energy "superpartner" boson**, and **vice versa**.
- **M-theory** unifies **five different rival versions** of string theory using an **eleven-dimensional spacetime**.

WHERE ARE THE EXTRA DIMENSIONS?

If **extra dimensions** exist, they are probably **compactified** – **curled up** on themselves at a **scale so small** that they **cannot be detected** – imagine that a ball or a tube were diminished to a point where they were indistinguishable from a **one-dimensional dot** or a **two-dimensional line** – it's all a matter of scale.

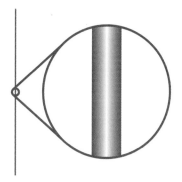

ROOTS OF RELATIVITY

The concept of relativity in physics has been around since its very beginnings – but it was only in the twentieth century that the idea began to exert a transformational effect on all other aspects of physics.

GALILEAN RELATIVITY

Galileo Galilei was the **first person** to state a **principle of relativity**. He recognized the importance of the idea that the **laws of physics** must be the **same for everyone** – or **otherwise no meaningful discussion of those laws could take place**:

- Every **experimenter's location** is in a **different state of motion** and/or **subject to other influences** that **distinguish them from everyone else**.
- Therefore **why should any particular experimenter** be "privileged" as the **only witness** to **physical events** as they **truly are**?
- Instead, **all experimenters** must be able to **perform experiments** that ultimately **lead to the same laws**.

THE MICHELSON–MORLEY EXPERIMENT

Questions of **relativity** became troubling in the late nineteenth century, when attempts to **measure Earth's movement** through the **aether** (the **medium supposedly responsible for transmitting electromagnetic waves**) **drew a blank**.

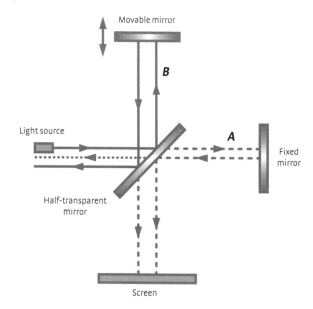

The **Michelson–Morley experiment**, conducted in **1887**, *should* have detected an **aether effect**:

- A **beam of light** is **split** and **sent along two paths** that **reflect back and forth multiple times**, at **right-angles** to each other. One beam moves **parallel to Earth's motion through space**, the other at **right-angles** to it.
- The two beams should therefore **travel at slightly different speeds** due to **Earth's motion** through the **aether**.
- The two **half-beams** are then **reunited** to create an **interference pattern**.
- The entire apparatus is **rotated**: according to predictions, this **should change** the **relative speeds** of the **split beams** and affect the **interference patterns**.
- However, **no changes were seen – light seems to move at the same speed regardless of direction**.

SPECIAL RELATIVITY

In order to resolve growing concerns about the classical Newtonian view of the Universe, Albert Einstein's special theory of relativity, published in 1905, rewrote the laws of physics from the ground up.

WHAT IS SPECIAL RELATIVITY?

Einstein's **"special" theory** is so called because it **applies to a limited range of situations** – those involving only **inertial frames of reference** that are **not undergoing acceleration**.

- **Frame of reference**: a **coordinate system** and the **set of reference points** needed to use it for purposes of **measurement** (such as an **arbitrary origin point** and a **unit of length**).

Einstein threw away traditional assumptions and instead looked again at the **laws of physics** with just **two postulates**:

- The laws of physics are **identical in all inertial frames of reference** (a restatement of **Galileo's principle of relativity**).
- The **speed of light in a vacuum** is the **same for all observers**, regardless of the **relative motion of source and observer**.

Guided by these two principles, he developed a number of **"thought experiments"** that allowed him to **consider what would happen in various situations**.

RELATIVITY OF SIMULTANEITY

One example of the **consequences of relativity** arises from considering **simultaneous events**. For example, imagine **two observers**, **one on a train platform** and **one in the middle of a passing train carriage**.

- At the moment the observers **pass each other**, a **flash of light** is **triggered in the middle of the carriage**.
- For the **moving observer**, the **two ends of the carriage** are **equidistant** and so **light** (moving at **constant speed**) **strikes the two ends simultaneously**.

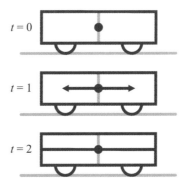

- For the **stationary observer**, light moving **towards the front of the carriage** has to **travel further to reach the end**, while light moving **backwards** finds the **end of the carriage coming to meet it**, and therefore **hits that end sooner**.

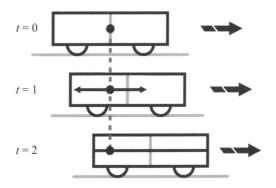

THE LORENTZ TRANSFORMATIONS

Einstein's equations of special relativity frequently feature a mathematical formula, originally derived by Hendrik Lorentz to predict the behaviour of objects moving through the aether.

THE LORENTZ FACTOR

Attempting to explain the **failure of the Michelson–Morley experiment**, several physicists came up with the idea that **movement through the aether caused objects to contract in the direction of motion**. This could **shorten the path taken by light** and **make up for the effect of the "aether wind" slowing it down**.

Lorentz calculated the effect in terms of a **Lorentz factor**, denoted as γ.

$$\gamma = 1\sqrt{(1 - (v^2/c^2))}$$

Although Lorentz's inspiration was **misguided, Einstein realized** that the **Lorentz factor** acts as **a guide** to the **behaviour of space and time** in situations where an **object** and an **observer** are **moving at very high speeds (close to the speed of light) relative to each other**.

When an object defined by coordinates x, y, z in **space**, and t in **time** moves at **velocity** v in the x **direction**, it experiences two **Lorentz transformations**:

$$x' = \gamma\,(x - vt)$$
$$t' = \gamma\,(t - vx/c^2)$$

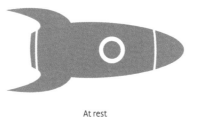

At rest

Because c is **huge compared to everyday speeds**, the Lorentz factor is very **close to 1** at all but the **highest "relativistic" speeds**. The transformations are therefore **indistinguishable** from their **expected values in classical physics**.

However, as the **relative speed approaches the speed of light**, v^2/c^2 becomes **significant** and the Lorentz factor **increases**:

Speed 0.1c	$v^2/c^2 = 0.01$	$\gamma = 1.005$
Speed 0.5c	$v^2/c^2 = 0.25$	$\gamma = 1.15$
Speed 0.9c	$v^2/c^2 = 0.81$	$\gamma = 2.29$
Speed 0.99c	$v^2/c^2 = 0.98$	$\gamma = 7.08$

Two key results of this are:

Lorentz contraction: the **moving object grows shorter in the direction of travel**.

Time dilation: the **object experiences** the **passage of time more slowly relative to an outside observer**.

MASS—ENERGY EQUIVALENCE

Alongside his first outline of general relativity, Einstein published another paper that questioned the widespread understanding of energy and mass. In it, he derived the most famous equation in physics.

INERTIA AND ENERGY

Special relativity places an **absolute upper speed limit on the Universe** – the **speed of light** itself. So what happens when an **object with mass attempts to approach that speed**? How can the **fixed speed of light** be **reconciled with conservation of energy and momentum**?

In a 1905 paper, Einstein explained that as a consequence of special relativity, **energy supplied to an object already moving at relativistic speeds increasingly boosts its *mass* rather than its velocity**. This allows the **object's energy** and **momentum to increase as they should**, but **limits further acceleration**.

The **energy** E, **mass** m, and **momentum** p of an object moving at **velocity** v are defined by:

$$E = \gamma\, E_0$$
$$m = \gamma\, m_0$$
$$p = \gamma\, mv$$

(where γ is the Lorentz factor).

E_0 is therefore the object's "**rest energy**" and m_0 its "**rest mass**" in situations where it is **stationary relative to an observer's frame of reference**. From these relationships, Einstein was able to show that **energy and mass are actually equivalent in all situations**, and are **linked by the famous equation**

$$E = mc^2$$

MASS, ENERGY, AND RADIOACTIVITY

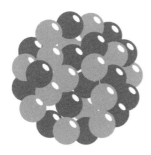

When scientists began to study the **different forms of radioactive emission** in the late nineteenth century, one **big question** was **where the energy to produce gamma rays came from**. How could such **energetic radiation** be **produced** by **changes in tiny, individual atoms**? Einstein's **equation** offered the solution – a **tiny fraction** of the **atom's mass** is **converted directly into energy** as the **atomic nucleus rearranges itself** and **neutrons transform into protons**.

Mass–energy of 1 neutron: 939.57 MeV

Mass–energy of 1 proton: 938.28 MeV

Mass–energy of 1 kilogram = 9×10^{16} joules

SPACETIME

One of the most powerful tools for thinking about special relativity was developed by Einstein's former tutor, Hermann Minkowski. In 1908 he showed that space and time must be treated as a unified whole: spacetime.

SPACETIME DIAGRAMS

Minkowski formulated a way of **treating events in space and time visually** using **diagrams**:

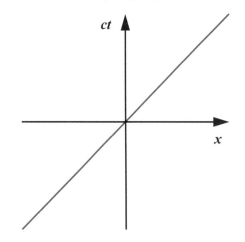

- The *x*-axis represents **movement** in a **single dimension of space**.
- The *y*-axis represents **movement through time** – labelling it *ct* (**distance** in terms of **light-travel time**), amounts to the same **thing**.
- **Objects** and **events** are depicted as **lines** across the diagram.
- With the **time axis** defined in terms of *ct*, **photons** – **travelling at the speed of light** – move through the diagram at **45-degree angles**.

SPACETIME AND GEOMETRY

One way of looking at the **difference** between the **two frames of reference** in the **moving-train example** is to see them as a **rotation of the axes** defining the **two frames of reference**. The **effects of these changes** can be **modelled** using the **rules of simple geometry**, and produce **results equivalent to the Lorentz factor**.

SIMULTANEITY REVISITED

Look again at the moving-train thought experiment (see p.157). The spacetime diagram for the observer on the train looks like this:

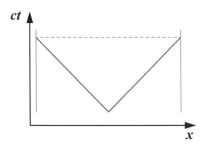

- **Photons of light move out from the centre at 45°.** The **ends of the train** do not **change their position**, so as "**events**" they form **vertical lines** that the **light strikes simultaneously**.

Now consider the diagram for the **observer on the platform**:

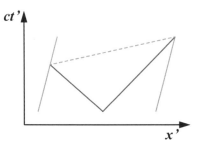

- The **light still moves out at 45°**, but now the **ends of the train are moving events** and are themselves **tilted at an angle**. The **light beam** now **intercepts one before the other**.
- NB: Angles are **exaggerated for clarity** – in reality the **effect at non-relativistic speeds** is **imperceptible**. But think about what happens at **relativistic speeds** as the **lines representing the walls tilt towards 45° themselves**.

GENERAL RELATIVITY

After the publication of special relativity, Einstein spent a decade developing a general theory that could be applied to accelerating, as well as non-accelerating, frames of reference.

THE EQUIVALENCE PRINCIPLE

Einstein's **key insight** on the **road to general relativity** came in 1907:

- A **stationary observer** on the **surface of the Earth** is not in an **inertial reference frame** – they are being **acted on** by a **steady downward accelerating force** – the **pull** of **gravity**.
- Therefore, in terms of the **forces acting** in a **particular frame of reference**, the **presence of** any **gravitational field** is **equivalent** to **constant uniform acceleration**.
- So the **effects** of **extreme gravitation** are **equivalent** to those of **relativistic motion**: they create **distortions of space and time** similar to those seen in **special relativity**.

EINSTEIN'S FIELD EQUATIONS

General relativity describes the **relationship of spacetime to gravitation** in a remarkably **simple mathematical form**:

$$R_{\mu\nu} - \frac{1}{2} R g_{\mu\nu} + \Lambda g_{\mu\nu} = \frac{8\pi G}{c^4} T_{\mu\nu}$$

The **mathematical meaning** of the various elements is beyond our scope, but $R_{\mu\nu}$ is the **Ricci curvature tensor**, R the **scalar curvature**, Λ (lambda) the **cosmological constant** driving the **expansion of space**, $g_{\mu\nu}$ the **metric tensor**, and $T_{\mu\nu}$ the **stress–energy tensor**. Within this maze of terminology, the **speed of light** c and **Newton's gravitational constant** G stand out.

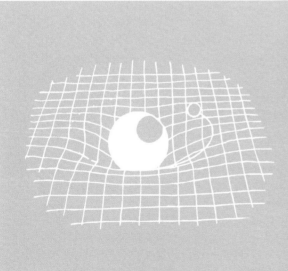

ANALOGIES OF RELATIVITY

The **"rubber sheet" model** is a common way of **visualizing the effects of general relativity** in more familiar terms. It **"discards"** one of the three **space dimensions** to **imagine space** as a **two-dimensional sheet. Large masses** warp the sheet **"downwards"** corresponding to **distortions** experienced by **objects** within the **warped region**.

The **rubber sheet analogy** should be **used carefully** – it's **not really representative** of what's going on. Another way of thinking about distortions of the **three space dimensions**, at least, is to consider them as **"pinching"**, like the neck of an hourglass, **around massive objects**.

GRAVITATIONAL LENSING

In 1919, measurements of stars near the Sun provided startling evidence for Einstein's ideas about the Universe. They demonstrated gravitational lensing – a phenomenon that has since become an important tool in modern astronomy.

SPACETIME LENSES

Because **large masses** in general relativity **distort nearby space and time** instead of simply **exerting a gravitational pull** on other masses, Einstein's theory predicts that **massless objects** such as **light rays** should be **affected by gravity**. This has **several results**:

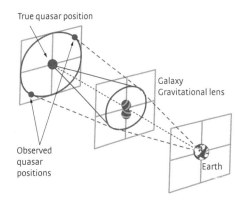

True quasar position

Galaxy
Gravitational lens

Observed
quasar
positions

Earth

- **Light** arriving at **Earth** from **distant objects** can come from **misleading directions**.
- **Images** of **large objects** can be **distorted**, often into **rings** or **arcs**.
- When objects are **aligned**, the **distortion** due to the **foreground object** can act as a **perfect lens**, redirecting **light rays** towards the **Earth** and creating an **unusually bright image** of the **background source**.

PROOF OF RELATIVITY

Published in 1915, **general relativity** remained an **intriguing theoretical idea** until 1919, when astronomer **Arthur Eddington** journeyed to the island of **Príncipe** off West Africa to observe a **total solar eclipse**. The eclipse **blocked out the Sun's light** and allowed him to **measure the positions of stars** whose **light** had passed **very close to the Sun**. The stars proved to be slightly **displaced** from their **expected locations**, proving that **gravitational lensing** was **real** and **Einstein** was **correct**.

LENSES AT WORK

Astronomers have found a number of **ingenious applications** for **gravitational lensing**. These include:

Lensing makes **small, distant galaxies** appear **far brighter**, bringing them into range for Earth-based and satellite telescopes.

Lensing is caused by **a foreground object's entire mass** (including invisible **dark matter** and **black holes**). By calculating the distortion effect of a gravitational lens, astronomers can work out **how much matter it contains**, and **how this is distributed**.

When **small Earthlike planets** pass in front of their **parent stars**, they can sometimes cause a **gradual brightening and fading of the star's light** due to "**microlensing**". This allows new planets to be **detected** and **weighed**.

GRAVITATIONAL WAVES

The last unproven prediction of general relativity, gravitational waves were undetectable for more than a century. Now that they have been found, they promise to open up new ways of investigating the Universe.

RIPPLES IN SPACE

Einstein's **field equations** of **general relativity** predict that **large moving masses** can create **distortions of spacetime** that **spread out** across the **Universe**. When the movements of masses are **periodic** (for instance when **two heavyweight stars orbit each other very rapidly**), the distortions can take the form of **periodic waves** rippling across the Universe.

Gravitational waves passing through Earth cause **minute periodic variations** in the **different dimensions of space**. These distortions are equivalent to a change by the **width of a proton** in the **4-kilometre length** of a **typical detector instrument**.

The distortions are measured by **laser interferometers** such as **LIGO** in the **USA** and **VIRGO** in **Italy**. These instruments **split a beam** of **finely tuned laser light**, sending it along **two perpendicular paths**, each of which involves **multiple reflections** equivalent to a **total distance of 1,120 km**. The **split beams** are then **recombined** – the **precise length** of each **travel path** determines how they **interfere with each other**, and the **passage of gravitational waves** creates **distinctive signals**.

Using interferometers in **different locations** and with **different orientations** allows scientists to detect the **direction** from which the waves are **passing through Earth**.

GRAVITATIONAL WAVE ASTRONOMY

- Gravitational waves detected so far come from the **violent mergers** of massive **dead stars** – the final moments in which **superdense neutron stars** or **black holes** trapped in **orbit** around each other **spiral together**, **collide**, and **coalesce**.

- Ultimately, astronomers hope to use gravitational waves to look through the **opaque wall** that surrounds the **visible Universe** (see p.167) and study the era of the **Big Bang** itself.

BLACK HOLES

Black holes are the strangest objects in the Universe – single points of infinite mass permitted by general relativity, sealed off from the Universe by a barrier zone with remarkable properties.

SINGULARITIES

The **equations of general relativity** give rise to the idea of **singularities** – **large concentrations of mass** at a **single point in spacetime**, where the **laws of physics** are **stretched** to their **extreme**.

In 1916, **Karl Schwarzschild** showed that a **singularity** must be **surrounded by a region** in which **gravity** is so **strong** – and **spacetime** so **distorted** – that **even light cannot escape**. This is known as the **event horizon**.

However, astronomers did not consider the possibility of **black holes** existing in the **physical Universe** until the **1950s**, and it was only the discovery of **degenerate neutron stars** (see p.139) in the **1960s** that confirmed that the **death of massive stars** could trigger a **collapse powerful enough** to create a singularity.

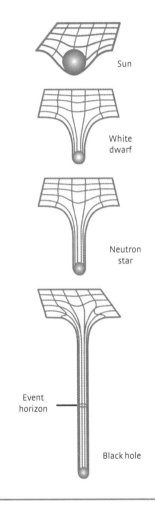

Sun

White dwarf

Neutron star

Event horizon

Black hole

AT THE EVENT HORIZON

Light approaching the **event horizon** is **stretched** to **longer wavelengths** and ultimately becomes **invisible**. **Matter falling inwards** experiences such strong **tidal forces** that it is **torn into its constituent atoms** in a process called **spaghettification**, and **heated** to **extreme temperatures**. This means that despite the **impossibility of radiation escaping from their surfaces**, they are **often surrounded by superheated discs of infalling matter** that **give off a wide range of radiations**.

BLACK HOLES IN THE UNIVERSE

Stellar-mass	Intermediate-mass	Supermassive black holes
Up to tens of solar masses	Hundreds of solar masses	Millions or billions of solar masses
Formed by the collapsing cores of the most massive stars	Formed by the merger of smaller black holes in the crowded hearts of star clusters	Formed from the runaway growth and merger of smaller black holes during galaxy formation

HAWKING RADIATION

Is the matter that falls into a black hole really trapped forever? Stephen Hawking proved that it is not, by discovering the strange radiation that bears his name.

PARTICLES ON THE BOUNDARY

- Thanks to the **time–energy uncertainty principle** (see p.133), **virtual particle pairs** are **constantly forming** and **annihilating** in space throughout the Universe.
- If a particle pair forms at the **very edge** of the **black hole**, one particle is **swallowed by the event horizon**, while the **other escapes into space**.
- The **surviving particle** is forced to become "**real**" – the **energy borrowed** from the **vacuum** cannot be **returned** through **annihilation**, so instead it is **stolen** from the **black hole** itself.
- **Viewed at a distance** from the event horizon, the result is a **perfect form** of **black-body radiation** (see p.62), with a **temperature inversely proportional** to the **mass of the black hole**: $T \propto 1/m$.

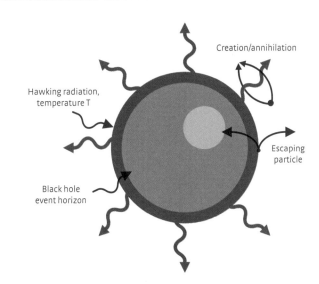

Creation/annihilation

Hawking radiation, temperature T

Escaping particle

Black hole event horizon

- Over time, **Hawking radiation** gradually **drains energy** and **mass** from the black hole (known as **evaporation**). Unless it receives **additional mass** from its surroundings, its **gravity** grows **steadily weaker** until eventually it **bursts apart** with a **flash of gamma rays**.
- **Stellar-mass black holes** absorb enough **energy** from their surroundings (even from the **cosmic microwave background** – see p.167) to **prevent their evaporation**.
- However, **small black holes** with a mass of 10^{11} to 10^{12} kg, which might have formed during the **Big Bang**, could **completely evaporate** in a few billion years, making their deaths **potentially observable** in today's Universe.

BLACK HOLES AND INFORMATION

According to **fundamental tenets** of **quantum physics**, **information** about the **quantum states of particles** cannot be **destroyed** – but **Hawking radiation**, when first identified, appeared to suggest the **opposite**: the **black-body radiation** should **theoretically** be **perfect**, with **properties** determined **entirely** by the **black hole's mass**. In the 1990s, however, **Hawking revised his ideas** to allow for **tiny quantum fluctuations** in the **event horizon itself** – a means of "**encoding**" the information from the **lost particles** so that it is **theoretically preserved**.

WORMHOLES AND TIME MACHINES

Alongside black holes, Einstein's field equations permit another bizarre spacetime structure called a wormhole. If they exist, wormholes could offer shortcuts across the Universe, and even allow the construction of time machines.

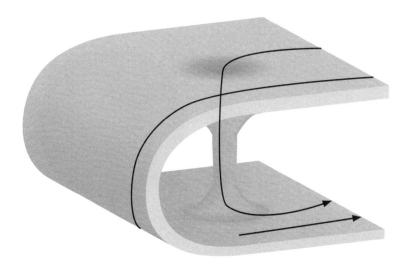

WORMHOLE GEOMETRY

Also known as an **Einstein–Rosen Bridge**, a **wormhole** is an **open tunnel** between **two distant regions of space and time**. Its structure is somewhat similar to a **black hole**, but instead of **pinching** into a **singularity**, the **warped spacetime** within it eventually **emerges** through a **second wormhole** in a **distant part** of **spacetime**.

A wormhole therefore offers a **theoretical cosmic shortcut** that might allow a **spacecraft** to **cross thousands of light-years** in a **relatively short period of time**, without needing to **break the light-speed limit**.

So far, astronomers have found **no evidence** of **natural wormholes** with **long-term stability**. It might one day be possible to construct an **artificial wormhole**, but **preventing the wormhole from collapsing** to a **singularity** would require **exotic matter** with hypothetical properties such as **negative mass**.

BUILDING A TIME MACHINE

If a wormhole could be **found** or **constructed**, it could also be used as the basis for a **time machine**. The principle relies on creating a **time difference** between the **two ends of the wormhole** while **bringing them close together** in space.

- Engineers from an advanced civilization **travel through** the wormhole.
- They create a method of **anchoring** the **far end** (for instance through its **gravitational attraction to a planet**) in a way that allows it to be **transported**.
- The far end of the wormhole is **dragged back** to the **origin** at **relativistic speeds**. **Time dilation** (see p.158) causes **time** to **travel more slowly** during its journey.
- By the time it is **returned** to the **origin point**, the **far end** has **slipped into the past**.
- **Travel through the wormhole** now involves a journey into the **past** or **future**. These "**time jumps**" are **unlimited in either direction** – but it's **impossible to travel back to before the time machine was created**.

THE LARGE-SCALE UNIVERSE

Cosmology is the branch of astronomy concerned with the large-scale structure, origin, and fate of the Universe. Its foundations lie in some remarkable discoveries made in the twentieth century.

SCALE OF THE UNIVERSE

Edwin Hubble, 1925
The existence of **external galaxies**, and the **vast scale** of **cosmic distance**, only became clear once **stars** whose **true luminosity** could be predicted were found in the mysterious "**spiral nebulae**". These showed that **most galaxies** were **millions of light-years from Earth**.

THE COSMIC TIME MACHINE

The **cosmic distance scale** is so **huge** that **even light** takes **millions of years** to reach Earth from **distant objects**. As **telescopes** have improved and **more distant objects** have **come into view**, our observations are showing us the Universe **as it looked** when it was **billions of years younger**.

COSMIC EXPANSION

Edwin Hubble, 1929
Doppler shifts in the **light** of **other galaxies** are consistently towards the **red end** of the **spectrum**, indicating that they are **moving away from us**. The **more distant** a galaxy is, the **faster its speed of retreat** (**Hubble's law**). This indicates that the **entire Universe is expanding** from an **earlier state** that was **denser** and therefore **hotter**.

MICROWAVE BACKGROUND

Arno Penzias and Robert Wilson, 1964
The sky is filled with **weak microwave radiation** coming from space in **all directions**, corresponding to a **background temperature** 2.7°C above **absolute zero**. This **Cosmic Microwave Background Radiation** (CMBR) is created by **light** from the era when the **Universe became transparent**, stretched into the **microwave region** during its **journey across space towards Earth**.

LIGHT-YEAR
One light-year is the **distance light travels during an Earth calendar year**:

1 ly = 9.5×10^{12} km / 5.9 million million miles

HUBBLE'S LAW
In the **local Universe**, the **speed of a galaxy's recession** is **related to its distance** by the **Hubble Constant**:

H_0 = c. 21.5 km/s per million light-years distance

THE BIG BANG THEORY

The Big Bang theory is the most successful model of cosmic evolution when it comes to describing where the Universe came from, and how it has evolved over 13.8 billion years.

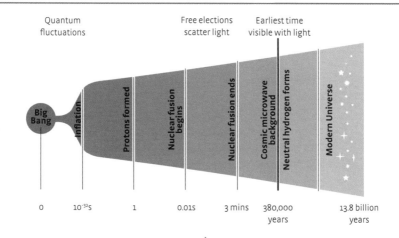

Quantum fluctuations · Free elections scatter light · Earliest time visible with light

Big Bang · Inflation · Protons formed · Nuclear fusion begins · Nuclear fusion ends · Cosmic microwave background · Neutral hydrogen forms · Modern Universe

0 · 10^{-32}s · 1 · 0.01s · 3 mins · 380,000 years · 13.8 billion years

DISCOVERING THE BIG BANG

Until the discovery of **cosmic expansion**, astronomers generally assumed that the Universe was **eternal**. The idea that it started out in a **compact, dense state** with much **higher temperatures** (a **"primeval atom"**) was first suggested by **Georges Lemaître** in 1931.

Advances in **nuclear** and **particle physics** soon showed that if the **expansion** was **wound back far enough**, **matter** would be **reduced to pure energy**.

The term "**Big Bang**" was initially an **insult** coined by **Fred Hoyle**, a supporter of the rival **"steady state" theory** in which **matter** was **continually created** to fill the gaps created by **expansion**.

In 1948, **Ralph Alpher** and **George Gamow** devised a theory of "**Big Bang nucleosynthesis**" to explain how the **pure energy** of the Big Bang gave rise to the **atoms** of **lightweight elements** that dominate the Universe.

Discovery of **Cosmic Microwave Background Radiation** in 1964 clinched the case for a Big Bang.

BIG BANG TIMELINE

Up to 10^{-43} seconds The Universe is so small that even the laws of quantum physics do not apply

10^{-36} seconds Broken symmetry between fundamental forces drives a sudden burst of expansion called inflation

Up to 10^{-6} seconds Energy is transformed into quark–antiquark pairs that mostly annihilate to release their energy again. However, a small excess of matter quarks survives

Up to 1 second Quarks can no longer form, but bind together into baryons

10 seconds Creation and annihilation of leptons and antileptons ceases, leaving a small excess of leptons. Photons now carry most of the energy in the Universe

2 to 20 minutes Baryons bind together to form lightweight atomic nuclei

Up to 380,000 years The Universe remains opaque – densely packed matter creates a fog that traps light bouncing between particles

380,000 years Temperatures drop to c.3000°C – cool enough for electrons and nuclei to combine in the first atoms. The density of matter drops and the Universe becomes transparent

Up to 150 million years Cosmic "dark age" until the first stars form

DARK MATTER

Measurements of our galaxy and others show that the normal matter we can see in the Universe accounts for just a small 15 per cent all the matter in the Universe. The rest is unknown, mysterious dark matter.

DISCOVERING DARK MATTER

Dark matter is not simply **dark** – it's also **entirely transparent**, refusing to **interact** with **electromagnetic radiation** in any way. It was discovered through **observations** on **two different scales**:

1. Galaxy clusters
Fritz Zwicky, 1933

Zwicky measured the galaxies in the **Coma galaxy cluster** and found that they were being **influenced by gravity** much **stronger** than the **number of visible galaxies in the cluster** would suggest. He estimated that **unseen dark matter** outweighed **visible matter** by 400 to 1.

2. Milky Way rotation
Vera Rubin, 1978

Rubin measured the **orbits of stars** in different parts of the **Milky Way** and found **variations** that could **not be explained** by the **galaxy's visible mass**. She concluded that the Milky Way contained five to ten times more **dark matter** than **visible matter**, especially in the **halo region** above and below its **star-rich spiral disc**.

- In recent decades, **improved observations** of **normal but non-luminous matter** (such as **cool dust clouds** that emit **infrared**) have confirmed that **dark matter** outweighs **visible matter** by about six to one.

SO WHAT IS IT?

Astronomers have investigated **two broad explanations** for dark matter:

- **Massive Compact Halo Objects** (MACHOs): Objects such as **black holes** and **stray planets**, **orbiting** in the **halo regions** around **galaxies** but **too small** and **faint** to **give their presence away**. **New observing techniques** have revealed some **MACHOs**, but also confirmed that they are **not present in large enough quantities** to **account for dark matter**.

- **Weakly Interacting Massive Particles** (WIMPs): New **elementary particles** that are **immune** to the **electromagnetic force**. **Neutrinos** (which were thought to be **massless** until 1998) have accounted for a **small proportion** of **dark matter**, but **other particles outside of the Standard Model** will be needed to explain it all.

DARK ENERGY

A revolutionary discovery in the late 1990s showed that the expansion of the Universe is not slowing down – as we might expect under the influence of gravity – but is instead speeding up due to a mysterious force called Dark Energy.

DISCOVERING DARK ENERGY

In the 1990s, **two teams** of astronomers set out to test an **ingenious new way of measuring** the **rate of cosmic expansion**, using **supernova explosions** in **distant galaxies**.

- **Type 1a supernovae** are a special form of **stellar explosion** triggered when a **white dwarf collapses into a neutron star**. They always **release** the **same** amount **of energy** and therefore have the **same luminosity**.

- This means the supernovae can be used as "**standard candles**" – objects with **known luminosity** whose **brightness** as **seen from Earth** confirms their **distance**.

- When the astronomers **compared** the **supernova brightness** with that **estimated** from the **redshifts** of their **host galaxies** according to **Hubble's law**, they found that the supernovae were **consistently fainter than expected**.

- This **discrepancy** can only be explained if the **rate of cosmic expansion** has **increased** over the **history of the Universe** – an effect attributed to a mysterious phenomenon called **dark energy**, which is now known to account for **68.3 per cent** of all the **Universe's energy content**.

FATES OF THE UNIVERSE

- The **balance** between **normal matter**, **dark matter** and **dark energy** plays a crucial role in determining the **eventual fate** of our Universe:

- If the Universe has **sufficient mass** and **gravity**, its **expansion** will gradually **slow** and eventually **reverse**, drawing it back to a "**Big Crunch**". This is currently thought to be very **unlikely**.

- In a Universe with **just the right amount of mass**, **expansion** will get **slower** and slower, but never **entirely stop**.

- In a Universe with **too little gravity** (or with significant **dark energy**), **expansion** will **continue forever**, with **galaxies** becoming more **widely spaced** and the **Universe cooling** in a "**Big Chill**" for eternity.

- If dark energy **increases significantly** over time, its **growth** may become **exponential**, affecting smaller and **smaller scales** until **galaxies**, **solar systems**, **planets**, and **atoms** are torn apart in a "**Big Rip**".

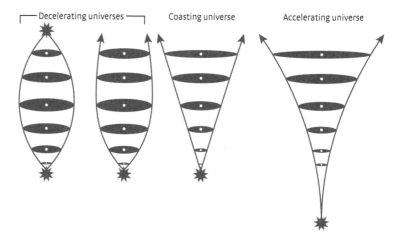

Decelerating universes   Coasting universe   Accelerating universe

THE ANTHROPIC PRINCIPLE

Is the presence of intelligent life in the Universe a cosmic coincidence, or does it have some deeper meaning? Physicists have drawn differing conclusions around the controversial idea known as the anthropic (human-centred) principle.

FINE-TUNED COSMOS

Martin Rees (*Just Six Numbers*) defined the Universe's **ability to support life** as **relying on** a number of **physical constants** that all seem to be suspiciously "**fine-tuned**":

- N, the **ratio of electromagnetic force** to **gravitational attraction** between **protons**.

- ε (epsilon): The **efficiency** of the **hydrogen–helium nuclear fusion reaction**.

- Ω (omega): The **density parameter** defining the **balance of gravity** against **cosmic expansion**.

- λ (lambda): The **cosmological constant**, defining the **amount of dark energy in the Universe**.

- Q: A **measure** of how easily **galaxies** can **form** and **remain stable**.

- D: The number of **spatial dimensions** in **spacetime**.

EVOLVING IDEA

In 1973, **Brandon Carter** suggested we should **not be surprised** to find that our Universe is **suitable for producing life**, since we are **here to observe it**. In other words, if the various **fine-tuned properties** had **values** very **different** from **those we measure today**, we **wouldn't be around to measure them**.

In 1986, **John Barrow** and **Frank Tipler** restated Carter's idea and defined it as the **weak anthropic principle**. They also define a **strong anthropic principle**: the idea that the **Universe has an imperative to produce life** – it really is fine-tuned, for one of **three possible reasons**:

- The Universe was **deliberately designed** by an **external agency** (be that a **deity** or an **alien intelligence**) in such a way as to **give rise to life**.
- The **presence of observers** is in some way **necessary** for the **Universe** to **come into being** (perhaps for reasons related to the **Copenhagen interpretation** of quantum physics.
- Our Universe is **one of many** in a **vast ensemble** that allows **all possible parameters** to be **explored** – we're only seeing a **small pocket** of an **overall multiverse**, to which the **weak anthropic principle** applies.

GLOSSARY

Alpha particle: A particle released from an atomic nucleus during radioactive alpha decay, consisting of two protons and two neutrons and equivalent to a helium nucleus.

Angular momentum: A property describing a spinning or circling object's tendency to continue its rotation, analogous to linear momentum and linked to the object's inertia and rate of rotation.

Atom: The smallest unit of matter that retains the properties of a chemical element (atoms were once thought to be indivisible but are now known to consist of subatomic particles).

Atomic mass: The mass of an atom in "atomic mass units" – equivalent to its total number of protons and neutrons. The atomic mass of an element is the weighted average of the masses of its various isotopes.

Atomic number: The number of protons within a specific atom, defining how many electrons are present in a neutral atom and which element it forms.

Beta particle: A particle released from an atomic nucleus during radioactive beta decay. Beta particles are usually electrons but occasionally positrons, released when a neutron transforms into a proton or vice versa.

Big Bang: An explosion in which the entire Universe, encompassing space, time, and all matter and energy, was created approximately 13.8 billion years ago.

Boson: A particle with zero or whole-number "spin" that displays unusual behaviour because it is immune to the Pauli exclusion principle. Gauge bosons are the bosons that act as force carriers in models of fundamental force interactions.

Current: A flow of electric charge through a conductor. Current is usually a flow of negatively charged electrons, but, by convention, it is treated as a movement of positive charge (flowing in the opposite direction to the electrons themselves).

Electromagnetic radiation: A wave generated by objects with changing electromagnetic fields, which consists of perpendicular electrical and magnetic waves regenerating one another as they move through space. Light, radio waves, X-rays, and gamma rays are all forms of electromagnetic radiation.

Electromagnetism: A fundamental force of nature that affects particles with electric charge, creating a repulsive force between similar charges, and an attractive force between opposite ones.

Electron: A lightweight elementary particle carrying a unit of negative electrical charge. Electrons are found in orbit around atomic nuclei, where they play a key role in chemistry, but they can also break free of atoms to act as charge carriers in electric current.

Endothermic process: A chemical or physical process that absorbs energy from its surroundings.

Entropy: A measure of the disorder in a system, and the amount of its energy that would be unavailable for harnessing by even an idealized heat engine.

Exothermic process: A chemical or physical process that releases excess energy into its surroundings.

Fermion: A particle with half-integer "spin" that is susceptible to the Pauli exclusion principle and therefore behaves in certain ways. All the elementary matter particles are fermions.

Force: An interaction that changes (or attempts to change) a body's motion. There are four fundamental forces in nature, each acting with different strengths on varied scales and between different types of particle.

Frame of reference: A system of coordinates used to measure the properties and behaviour of objects. The principle of relativity states that measurements taken in any non-accelerating frame of reference will always produce the same result, but observations made between frames of reference that are in relative motion can differ significantly.

Fundamental force: Any of the four forces that produce interactions between matter particles – electromagnetism, the strong and weak nuclear interactions, and gravitation.

Gamma radiation: The highest-energy, shortest-wavelength form of radiation, emitted by the nucleus during radioactive decay.

Gravitation: An attraction that acts between all objects with mass and creates an accelerating force (gravity) between them. According to general relativity, the effects of gravitation arise from the distortion of spacetime around massive objects.

Heat engine: Any device that makes use of the energy contained within heated molecules to do mechanical work, such as driving a piston.

Inertia: The innate tendency of objects with mass to resist forces that attempt to change their motion.

Inverse square law: A decrease in the influence of a property or force that decreases in inverse proportion to the increasing square of distance from the source. Inverse-square relationships are commonly found in physics, reflecting the way that many forces become more thinly spread across space.

Ion: An atom-like particle with a net electric charge due to an imbalance between the number of protons in its nucleus and the number of electrons in its outer layers – an excess of electrons creates a negative ion, while a shortfall creates a positive one.

Ionizing radiation: A catch-all term for particles and gamma rays emitted during the radioactive decay of unstable radioisotopes into more stable forms. The energy of the particles involved frequently ionizes other materials in their surroundings.

Isotope: Any one of several atoms that may have the same number of protons (and hence form the same chemical element) but that differ in the number of neutrons and hence their mass.

Lepton: A matter particle, such as an electron or neutrino, that is susceptible to weak, but not strong, nuclear interactions.

Light: A form of electromagnetic radiation with wavelengths between 400 and 700 nanometres (billionths of a metre), which human eyes have evolved to see.

Magnetic moment: A property related to a particle's charge and its spin or angular momentum, which determines the strength of its magnetic field.

Magnetism: An aspect of the electromagnetic force created by electrically charged particles in motion or rotation.

Mass: A property reflecting an object's inertia and the amount of matter it contains. Mass is created by the interaction of matter particles with the all-pervading Higgs Field.

Momentum: A property reflecting the force required to change an object's velocity, found from an object's mass multiplied by its velocity in a certain direction. When particles in a system collide, their combined momentum remains the same.

Neutron: A subatomic particle with no electric charge but similar (though not identical) mass to the proton, found in all atomic nuclei except the simplest form of hydrogen.

Nuclear fusion: A natural process that takes place at high temperatures and pressures in the cores of stars, joining together nuclei of light elements such as hydrogen to create heavier ones and release energy.

Nuclear fission: A process (usually artificial) of splitting the nuclei of heavy atoms to produce lighter ones and releasing energy as a by-product.

Nucleon: A catch-all term for the protons and neutrons of the atomic nucleus. An atom's atomic mass is a measure of the number of nucleons it contains.

Orbit: The elliptical path followed by one body around another due to the force of gravity.

Pauli exclusion principle: A law that determines much of the structure of matter by preventing fermion particles from taking on entirely identical "states" within a system such as an atom.

Photon: A small burst of electromagnetic radiation that can display behaviour similar to both waves and particles.

Potential difference: A measure of the potential energy difference between two points in an electric field, measured in volts.

Potential energy: The energy a particle or object possesses due to its location within a force field.

Proton: A positively charged subatomic particle with substantial mass, composed of quarks and found in the atomic nucleus.

Quantum: The smallest amount of a particular physical property that can take part in a physical interaction. Many physical properties, such as the energy carried by light waves and electrons, turn out to be quantized on the very smallest scales.

Quark: A fermion or matter particle that is susceptible to both the strong and weak nuclear interactions. Quarks have substantial mass, and bind together in pairs or triplets. Protons and neutrons are made from triplets of the two most common quarks.

Radioisotope: An isotope of an atom that is unstable, often due to having significantly more neutrons than protons in its nucleus) and is therefore prone to radioactive decay.

Spin: A property found in subatomic particles that is analogous (but not identical) to the angular momentum of larger bodies, and that governs some fundamental aspects of their behaviour.

Strong interaction: A fundamental force of nature that is very powerful but acts on very small scales, binding quarks together inside nucleons, and binding nucleons more weakly in the atomic nucleus.

Uncertainty principle: A law operating on the smallest quantum scales to prevent certain pairs of properties (such as position and momentum, or time and energy) being determined simultaneously with absolute precision.

Vacuum tube: A device that generates electrical effects by transmitting electrons through a vacuum, between two plates with a large potential difference between them.

Velocity: A measure of a body's speed (rate of motion) in a specific direction.

Virtual particle: A particle that can briefly wink in and out of existence thanks to the uncertainty principle linking time and energy. The exchange of virtual bosons is key to the operation of fundamental forces.

Wave: A travelling disturbance that propagates energy from one place to another, usually by triggering oscillations of a transmitting medium.

Weak interaction: A fundamental force of nature acting on the tiny scales of the atomic nucleus. The weak interaction affects all matter particles and is responsible for radioactive beta decay.

Weight: The reaction force that a body with mass exerts on anything that prevents its free acceleration through a gravitational field. Weight is properly measured in newtons and should not be confused with mass.

FURTHER READING

Jim Al-Khalili, *Quantum: A Guide for the Perplexed*

Marcus Chown, *Infinity in the Palm of Your Hand*

Albert Einstein, *Relativity: The Special and General Theory*

Richard Feynman, *Six Easy Pieces: Fundamental of Physics Explained*

James Gleick, *Isaac Newton*

John Gribbin, *In Search of Schrödinger's Cat*

Stephen Hawking, *A Brief History of Time*

John L. Heilbron, *Galileo*

Basil Mahon and Nancy Forbes, *Faraday, Maxwell and the Electromagnetic Field: How Two Men Revolutionized Physics*

Roger Penrose, *The Road to Reality: A Complete Guide to the Laws of the Universe*

Martin Rees, *Just Six Numbers: The Deep Forces that Shape the Universe*

Carlo Rovelli, *Seven Brief Lessons on Physics*